Security in Wireless Ad Hoc and Sensor Networks

Security in Wireless Ad Hoc and Sensor Networks

Erdal Çayırcı

NATO Joint Warfare Centre, Norway

Chunming Rong

University of Stavanger, Norway

A John Wiley and Sons, Ltd, Publication

Registered office
John Wiley & Sons Ltd, The Atrium, Southern Gate, Chichester, West Sussex,
PO19 8SQ, United Kingdom

For details of our global editorial offices, for customer services and for information about how to apply
for permission to reuse the copyright material in this book please see our website at www.wiley.com.

Library of Congress Cataloging-in-Publication Data

Çayırcı, E. (Erdal)
   Security in wireless ad hoc and sensor networks / Erdal Çayırcı, Chunming Rong.
      p.   cm.
   Includes bibliographical references and index.
   ISBN 978-0-470-02748-6 (cloth)
   1. Ad hoc networks (Computer networks)—Security measures.   2. Ad hoc networks
   (Computer networks)—Security measures.   I. Rong, Chunming.   II. Title.
   TK5105.59.C387 2009
   005.8—dc22

                                                                        2008041824

A catalogue record for this book is available from the British Library.

ISBN 978-0-470-02748-6 (H/B)

Set in 10/12pt Times by Integra Software Services Pvt. Ltd, Pondicherry, India
Printed in Great Britain by CPI Antony Rowe, Chippenham, Wiltshire

To Tülin and Ertuğ
Erdal Çayırcı

To Zhihua and Helena
Chunming Rong

Contents

About the Authors

Erdal Çayırcı graduated from the Army Academy in 1986 and from the Royal Military Academy, Sandhurst in 1989. He received his MS degree from the Middle East Technical University and a PhD from Bogazici University both in computer engineering in 1995

and 2000, respectively. He retired from the army when he was a colonel in 2005. He was an Associate Professor at Istanbul Technical University, Yeditepe University and the Naval Sciences and Engineering Institute between 2001 and 2005. Also in 2001, he was a visiting researcher for the Broadband and Wireless Networking Laboratory and a visiting lecturer at the School of Electrical and Computer Engineering, Georgia Institute of Technology. He is currently Chief, CAX Support Branch in NATO's Joint Warfare Center in Stavanger, Norway, and also a professor in the Electrical and Computer Engineering Department of the University of Stavanger. His research interests include military constructive simulation, sensor networks, mobile communications and tactical communications.

Professor Çayırcı has acted as an editor of the journals *IEEE Transactions on Mobile Computing, Ad Hoc Networks* and *Wireless Networks*, and has guest edited four special issues of *Computer Networks*, *Ad Hoc Networks* and *Special Topics in Mobile Networking and Applications (MONET)*.

He received the 2002 IEEE Communications Society Best Tutorial Paper Award for his paper entitled 'A Survey on Sensor Networks' published in the *IEEE Communications Magazine* in August 2002, the Fikri Gayret Award from the Turkish Chief of General Staff in 2003, the Innovation of the Year Award from the Turkish Navy in 2005 and the Excellence Award in ITEC in 2006.

Chunming Rong received his bachelors, masters and PhD degrees in Computer Science from the University of Bergen in Norway in 1993, 1995 and 1998, respectively. In 1995–1998, he was a research fellow at the University of Bergen. In 2001–2003, he was a post-doctoral researcher funded by Simula Research Laboratory. Currently, he is a Professor and chair of the computer science section at the University of Stavanger. He has also served as an adjunct Professor at the University Graduate Centre, University of Oslo, since 2005.

Professor Rong was given the ConocoPhilips Communication Award (Norway) in 2007. His paper *'New Infinite Families of 3-Designs from Preparata Codes over Z_4'* was awarded Editor's Choice in *Discrete Mathematics* in 1999.

He is an associate editor of the *International Journal of Computer Science & Applications (IJCSA)* and served on the editorial board of the *International Journal of Mobile Communications (IJMC)* between 2003 and 2006. For the IEEE International Symposium on Security in Networks and Distributed Systems (SSNDS), he was program chair in 2007 in Canada and general chair in 2008. For the International Conference on Autonomic and Trusted Computing (ATC), he was award chair in 2007 in Hong Kong and general chair in 2008 in Norway. For the International Conference on Ubiquitous Intelligence and Computing (UIC), he was general chair in 2008 in Norway.

Professor Rong was chairman of the board of the Foundation of the Norwegian Computer Science Conference (NIK) from 2005–2007, a board member of the Norwegian Information Security Network (NISNet) from 2007–2011 and a member of the Norwegian Informatics Council (Nasjonalt fagråd for informatikk). He has also been a member of the board for the 'ICT Security and Vulnerability (IKT-SoS)' program at the Research Council of Norway. He also currently serves in the workgroup for Information Security in Integrated Operation at the Norwegian Oil Industry Association (OLF).

As project manager, he has received grants from the Research Council of Norway for the projects 'Integration of Data Processing in Oil and Gas Drilling and Completion' for 2008–2010, 'Secure and Reliable Wireless and Ad Hoc Communications (SWACOM)' for 2006–2009 and 'Integrated IP-based Services for Smart Home Environment (IS-Home)' for 2007–2010. The Norwegian Information Security Network (NISNet) also receives annual funding from the Research Council of Norway.

His research interests include computer and network security, wireless communications, cryptography, identity management, electronic payment, coding theory and semantic web technology.

Preface

Ad hoc networks formed by randomly deployed self-organizing wireless nodes have a wide range of applications, such as tactical communications, disaster relief operations and temporary networking in sparsely populated areas, and therefore they have been studied extensively for two decades. More recently, sensor networks have attracted interest from the research community and industry. They are more energy constrained and scalable ad hoc networks. Another form of ad hoc network, namely mesh networks, is aimed at application areas such as infrastructureless network scenarios for developing regions, low-cost multihop wireless backhaul connections and community wireless networks. Characteristics such as wireless access, mobility, rapid and random deployment make these kinds of network a very challenging field for security. Security is also a key issue in making many ad hoc application scenarios practical. Although security for these networks has been studied extensively for more than a decade, there are still many challenges waiting for better solutions. Therefore, many researchers and engineers from both academia and industry continue working on this topic.

The book is designed for a 14–18 week (three hours a week) graduate course in computer engineering, communications engineering, electrical engineering or computer science. Prerequisite knowledge on computer networking is required. The book is self-contained with regard to wireless ad hoc networking issues and introduces the security-related aspects of wireless ad hoc, sensor and mesh networks, providing advanced information on security issues for this domain. The book may also be used as a reference work and readers are also likely to include engineers, either from networking or security fields, in industry or the military, who wish to perform protocol, network and security system design and implementation tasks for wireless ad hoc, sensor and mesh networks.

The book has two parts. The first introduces fundamentals and key issues related to wireless ad hoc networking. In this part, security-related issues – the issues referred to in the second part – are emphasized. In the second part, security attacks and counter measures in wireless ad hoc, sensor and mesh networks are elaborated upon. There is also a very short chapter about information operations and electronic warfare (EW) where the related terminology is introduced. Available standards about the related topics are also briefly presented in the last chapter.

Acknowledgements

We would like to thank our PhD students Dr Turgay Karlidere, Yan Liang and Son Thanh Nguyen. Turgay contributed Chapter 4 about Medium Access Control and provided us with the sections about the standards in that chapter. Son is the author of Chapter 15, which is about the standards on data security. Yan helped us edit two chapters.

Section 10.3 was adapted from Hegland *et al.* (2006), a paper we co-authored. We would like to thank our fellow authors of this paper, A.M. Hegland, E. Winjum, S.F. Mjølsnes, Ø Kure and P. Spilling, for letting us include the paper in this book.

List of Acronyms

AAA	Authentication authorization accounting
ACQUIRE	Active query forwarding in sensor networks
ADC	Analog-to-digital conversion
AES	Advanced encryption standard
AH	Authentication header
AKA	Auxiliary key agreement
AM	Access mesh
AOA	Angle of arrival
AODV	Ad hoc on-demand distance vector routing
ARAN	Authenticated routing for ad hoc networking
ARIADNE	On-demand secure ad hoc routing
BACNet	Building automation and control network
BAN	Body area network
BEC	Backward error correction
BGP	Border gateway protocol
BM	Backbone mesh
BWA	Broadband wireless access
C2	Command and control
C4ISR	Command, control, communications, computer, intelligence, surveillance, reconnaissance
C4ISRT	Command, control, communications, computer, intelligence, surveillance, reconnaissance, targeting
CA	Certificate authority
CA	Collision avoidance
CBRN	Chemical, biological, radiological and nuclear
CCMP	Counter mode with cipher block chaining message authentication code protocol
CD	Collision detection
CDMA	Code division multiple access
COMINT	Communications intelligence
CRC	Cyclic redundancy check
CRL	Certificate revocation list
CRS	Charging and rewarding scheme

CSMA	Carrier sense multiple access
CTS	Clear to send
DADMA	Data aggregation and dilution by modulus addressing
DCF	Distributed coordination function
DCMD	Detecting and correcting malicious data
DES	Data encryption standard
DISN	Defense information system
DLL	Data link layer
DoS	Denial of service
DS	Direct sequence
DSR	Dynamic source routing
EAP	Extensible authentication protocol
ECM	Electronic counter measure
EGP	Exterior gateway protocol
ELINT	Electronic intelligence
EMP	Electromagnetic pulse
EPM	Electronic protection measure
ESM	Electronic support measure
ESP	Encapsulated security payload
ESRT	Event-to-sink reliable transport
EW	Electronic warfare
FDD	Frequency division duplexing
FDM	Frequency division multiplexing
FDMA	Frequency division multiple access
FEC	Forward error control
FH	Frequency hopping
GPS	Global positioning system
HCI	Human–computer interaction
HMAC	Hash message authentication code
IBC	Identity-based cryptography
IBE	Identity-based encryption
ICMP	Internet control message protocol
IDS	Intrusion detection system
IEEE	International Electrical and Electronics Engineers
IF	Intermediate frequency
IGP	Interior gateway protocol
IHL	IP header length
IKA	Initial key agreement
INSENS	Intrusion-tolerant routing in wireless sensor networks
IP	Internet protocol

IrDA	Infrared data association
IS–IS	Intermediate system–intermediate system
ISM	Industrial, scientific and medical
IV	Initialization vector
LA	Location area
LAS	Local area subsystem
LDAP	Lightweight directory access protocol
LEACH	Low-energy adaptive clustering hierarchy
MAC	Medium access control
MAC	Message authentication code
MACA	Multiple access with collision avoidance
MACAW	Multiple access with collision avoidance, wireless
MARQ	Mobility-assisted resolution of queries
MD	Message digest
MIC	Message integrity code
MIMO	Multiple input, multiple output
MMSE	Minimum mean square estimation
MPDU	MAC protocol data unit
MR	Mobile radio
MS	Mobile subsystem
MT	Mobile terminal
NATO	North Atlantic Treaty Organization
NAV	Network allocation vector
OSI	Open system interconnection
OSPF	Open shortest path first
PAMR	Power-aware many-to-many routing
PC	Personal computer
PCF	Point coordination function
PCM	Pulse code modulation
PDA	Personal digital assistant
PKC	Public key cryptography
PKG	Private key generator
PKI	Public key infrastructure
PRMA	Packet reservation multiple access
PSFQ	Pump slowly, fetch quickly
PSK	Preshared key
QAM	Quadrature amplitude modulation
QOS	Quality of service
QPSK	Quadrature phase shift keying
QRS	Quarantine region scheme

RADIUS   Remote authentication dial in user service
RAP   Radio access point
RBS   Reference broadcast synchronizations
RC   Rivest cipher
RIP   Routing information protocol
RMST   Reliable multisegment transport
RSA   Ron Rivest, Adi Shamir, Len Adleman
RSN   Robust security network
RSS   Received signal strength
RTS   Request to send

SA   Security association
SAODV   Secure ad hoc on-demand distance vector
SHA   Secure hash algorithm
SIGINT   Signal intelligence
SLSP   Secure link state routing protocol
SMCS   System management and control subsystem
SN   Sequence number
SNDV   Sensor network database view
SNEP   Sensor network encryption protocol
SNR   Signal-to-noise ratio
SPAAR   Secure position-aided ad hoc routing
SPI   Sequence parameter index
SPIN   Sensor protocols for information via negotiation
SPINS   Security protocols for sensor networks
SOHO   Small office, home office
SQTL   Sensor query and tasking language

TACOMS   Post-2000 tactical communications
TCP   Transmission control protocol
TDD   Time division duplexing
TDM   Time division multiplexing
TDMA   Time division multiple access
TDOA   Time difference of arrival
TESLA   Timed efficient stream loss-tolerant authentication
TKIP   Temporal key integrity protocol
TOA   Time of arrival
TPSN   Timing-sync protocol for sensor networks
TRANS   Trust routing for location-aware sensor networks

WAS   Wide area subsystem
WASM   Wireless ad hoc, sensor and mesh network
WEP   Wired equivalent privacy
Wi-Fi   Wireless fidelity
WiMAX   Worldwide interoperability for microwave access

WLAN	Wireless local area network
WMAN	Wireless Metropolitan Area Networks
WMN	Wireless mesh network
WPA	Wi-Fi protected access
WPAN	Wireless personal area network
WSAN	Wireless sensor and actuator network

Part One

Wireless Ad Hoc, Sensor and Mesh Networking

1

Introduction

Although the wireless medium has limited spectrum and additional constraints when compared to guided media, it provides the only means of mobile communication. In addition, more effective usage of the limited spectrum and advanced physical/data link layer protocols enable broadband communications and integrated services over the limited wireless spectrum. Moreover, random and rapid deployment of a large number of tetherless nodes is possible through wireless ad hoc networking, which is a technology with a wide range of applications such as tactical communications, disaster relief operations and temporary networking in areas that are not densely populated. As a result, the use of wireless ad hoc networking has become pervasive. However, wireless ad hoc networking also introduces additional security challenges on top of those that exist for tethered networking:

- the wireless broadcast medium is easier to tap than guided media;
- the wireless medium has limited capacity and therefore requires more efficient schemes with less overhead;
- the self-forming, self-organization and self-healing algorithms required for ad hoc networking, and the schemes that tackle challenges such as hidden and exposed terminals, may be targeted to design sophisticated security attacks;
- the wireless medium is more susceptible to jamming and other denial-of-service attacks.

Wireless sensor and actuator networks (WSANs) are based on the random deployment of a large number of tiny sensor nodes and actuators into or very close to the phenomenon to be observed. They facilitate many application areas such as tactical surveillance by military unattended sensor networks, elderly and patient monitoring by body area networks (BANs) and building automation by building automation and control networks (BACnets). They are, in essence, ad hoc networks with additional and more stringent constraints. They need to be more energy efficient and scalable than conventional ad hoc networks, which exacerbates the security challenges. The security schemes for WSANs should require less computational power and memory because sensor nodes are tiny and have more limited capacity than the typical ad hoc network nodes such as a personal digital assistant (PDA) or a laptop computer.

Security in Wireless Ad Hoc and Sensor Networks Erdal Çayırcı and Chunming Rong
© 2009 John Wiley & Sons, Ltd

The wireless mesh network (WMN) is another member of the ad hoc network domain. WMNs enable application areas such as infrastructureless networks for developing regions, low-cost multihop wireless backhaul connections and community wireless networks. Actually, ad hoc networks can be considered a subset of WMNs because WMNs also provide a wireless backbone for working other mesh, ad hoc or infrastructure-based networks such as the Internet, IEEE 802.11, IEEE 802.15, IEEE 802.16, cellular, wireless sensor, wireless fidelity (Wi-Fi), worldwide interoperability for microwave access (WiMAX) and WiMedia networks. Lack of central authority and the availability of various access technologies to access the network make WMNs a more challenging domain in terms of security.

1.1 Information Security

In order to make the above and many more ad hoc application scenarios practical, they need to be secured against attacks. A security attack is an attempt to compromise the security of information owned by others (RFC 2828). We classify and examine all the security attacks designed to target ad hoc networks in the second part of this book. Security services are needed to defend against these attacks and to ensure the security of the information; these services can be categorized into two broad classes, namely *communications security* and *computer security* (Figure 1.1). Communications security defends against passive or active attacks through communication links or unintentional emanations. It ensures that communication services continue with the required level of quality and that classified data or information cannot be derived or captured from communications by an unauthorized node. Computer security ensures the security of computer hardware and software. It detects when a node or host is compromised, and recovers that specific node or host from the attack.

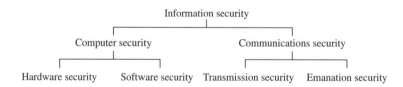

Figure 1.1   Information security

1.1.1 Computer Security

Host computers or network nodes can be attacked physically, and certain hardware components may be replaced, damaged or put out of service. Moreover, hardware can be infected by microbes which eat electronic components. Hardware security is designed to prevent, detect and repair these kinds of physical attack.

Another form of attack against computer hardware, especially in a battlefield, may be carried out by electromagnetic pulse (EMP) weapons. Portable systems that can emit EMPs are now available and critical hardware can be built to resist such attacks. Note that EMPs not only damage the software but also burn the hardware.

Viruses, worms and Trojan horses are examples of techniques used to attack software. They are programs that can infect an adversary's computer. A virus can duplicate itself many times and is generally designed to cause damage to the attacked software. A Trojan horse is a means of gaining unauthorized access to an attacked system. Finally, a worm copies itself from one node to another in the network and consumes resources such as computational power and memory. There are various software packages to eliminate these kinds of threat. However, most of them can detect only known viruses, worms and Trojan horses. Therefore, they rely heavily on the availability of the signature related to the attacking program.

1.1.2 Communications Security

An important part of communications security is transmission security, which is designed to prevent classified data in transmission being disclosed to unauthorized recipients, to defend against attacks that compromise a computer via communications links and to ensure that communication services are not hindered by malicious attacks. Note that once a computer is compromised or infected by a virus via a network connection, counteracting this falls under the remit of computer security. However, a compromised node can be used to attack a network from inside, and securing a network from an internal attack should also be considered within the scope of transmission security.

Emanation security is another important part of communications security. Computers may unintentionally transmit through Van Eck radiation or conduction. Every piece of electronic equipment, for example a computer, photocopier, printer, telephone, etc, radiates electromagnetic waves called *Van Eck radiation*. It is possible to receive this radiation and fuse the screen shots, key strokes and copied documents from a distance. Conduction of classified data through power lines, metal ducts, water pipes, wires and cables that are close to the media that carry or store these data is also possible. Emanation security aims to prevent this, and this field has been studied extensively by the military. The system is called TEMPEST, which is not an abbreviation but a word made up for military purposes.

1.2 Scope of the Book

In this book, we focus on transmission security for wireless ad hoc, sensor and mesh networks (WASMs). Transmission security has also been called *network security* or *Internet security* by other authors. All these terms are synonymous for us and incorporate all information security issues except for computer security and emanation security, explained in the previous subsection. Although our focus is on transmission security, we refer to both computer security and emanation security as they become relevant, because Van Eck radiation and EMP in particular are important issues for the military, and one of the most important application areas for WASMs is tactical field. Note that the utmost care has been taken not to disclose any classified information in this book. Everything explained in the book is generic, i.e. it does not reflect any specific tactical system but general requirements and design principles, all of which are already in the public domain.

The book elaborates all the known security weaknesses of WASMs and explains the potential threats and attacks that can capitalize on these weaknesses. We examine how to provide the following security services in the presence of these potential threats and attacks:

- authentication to ensure that a message is from the claimed source;
- access control to prevent the unauthorized use of network resources;
- data confidentiality to ensure that classified data cannot be disclosed by an unauthorized recipient;
- data integrity to ensure that a message is not modified during transmission;
- nonrepudiation to ensure that both the source and the destination are as specified in the message.

Other security-related issues such as the following are also within the scope of this book:

- defense against denial-of-service (DOS) attacks;
- reliable end-to-end services in a hostile environment;
- secure routing;
- measures to prevent the misuse of limited resources;
- measures to reduce the cost, i.e. computational, memory and power consumption, of security schemes.

1.3 Structure of the Book

This book is designed as a self-contained textbook for both academicians and practitioners. Therefore, it has two parts. It first introduces fundamentals and key issues related to WASMs. Although the primary goal of the first part is to provide the reader with the requisite background knowledge, we also highlight and explain the facts that have an impact on information security in this part. In the second part, we present advanced information on security for WASMs.

The first part of the book consists of seven chapters including the Introduction. The second chapter explains WASMs and the application areas. We also provide a subsection about tactical communications in this chapter because it is one of the most important application areas and has some special features that impact on security. Then we examine the factors influencing the design of the networks in our domain and how they impact on security schemes.

The fundamentals about the wireless medium are explained in Chapter 3, where the requisite knowledge for the physical layer is provided. The propagation environment, modulation, wireless channel impairments, jamming and the security considerations related to these issues are elaborated in this chapter.

Chapters 4, 5 and 6 explain the challenges and solutions in the data link, network and transport layers respectively. This layered approach is designed to provide interoperability and reusability of the protocols and schemes. However, it does not always lead to the optimal solutions. Therefore, transparency from the lower layer details, interoperability and reusability of protocols may be achieved at the expense of a more costly protocol stack compared to the cross-layer design. Since WASMs introduce some very stringent constraints; cross-layer protocols are common in our domain.

In Chapter 7, challenges specific to WASMs, such as node localization, time synchronization, addressing, coverage, mobility and resource management, are introduced together with the solutions available in the literature. Again, the security implications of the schemes that tackle these challenges are presented.

The second part of the book consists of eight chapters and starts with a detailed discussion of security attacks in WASMs. Then the cryptographic primitives are explained in Chapter 9. Challenges and solutions related to the basic issues such as bootstrapping, key distribution and integrity are covered in Chapter 10. Chapter 11 is about the challenges and solutions related to protection such as privacy, anonymity, intrusion detection, traffic analysis, access control, tamper resilience, availability and plausibility.

The self-forming, self-organization and self-healing features of ad hoc networks create new challenges for information security. These challenges and solutions related to secure routing are provided in Chapter 12, which is followed by another chapter about challenges and solutions specific to WASMs.

We also provide a short introductory chapter about information operations and electronic warfare and how this relates to security in WASMs in Chapter 14. Note that electronic warfare is a huge topic alone, and our intention is to provide only introductory coverage. Our final chapter is about the standards related to the security in wireless networks.

1.4 Electronic Resources for the Book

The book has a website at http://www.securityinadhoc.net/. At this site you can access the following:

- **Slides:** Powerpoint slides to complement the content of the book are available at this link. The slides are designed for a 14-week course where the book provides the text. They are kept updated.
- **Updates table:** any updates and corrections to the book are listed in this table. If you have any comments or suggestions on how to improve our book, please email them to us by using the 'contact us' link on the website.
- **The interest group:** this link provides you with a user name and password to access the interest group for the content of the book.
- **Useful links:** links to useful websites are provided here.

1.5 Review Questions

1.1 What is information security? How can you categorize it?

1.2 What is Van Eck radiation?

1.3 What is emanation through conduction? How can it be prevented?

1.4 What are the differences between a virus, a worm and a Trojan horse?

1.5 What is an electromagnetic pulse attack?

1.6 What is a security attack?

1.7 What is a denial-of-service attack?

1.8 What are the security services?

2

Wireless Ad Hoc, Sensor and Mesh Networks

Wireless networking paradigms can be categorized broadly into two classes: *wireless ad hoc* and *cellular* networking. The existence of a fixed infrastructure is the main difference between these two classes (Figure 2.1). In the ad hoc networking paradigm there is no fixed infrastructure and packets are delivered to their destinations through wireless multihop connectivity. Nodes often act not only as hosts but also as routers, relaying the traffic by other nodes. The topology of an ad hoc network can change because nodes may not be fixed or, equally, they may fail. In the cellular networking paradigm, mobile terminals reach an access point via a single-hop wireless link. There is a fixed infrastructure that connects the access points to each other. Therefore, ad hoc and cellular networks are often called *infrastructure-less* and *infrastructured* networks respectively. The basic characteristics of ad hoc and cellular networks are compared in Table 2.1.

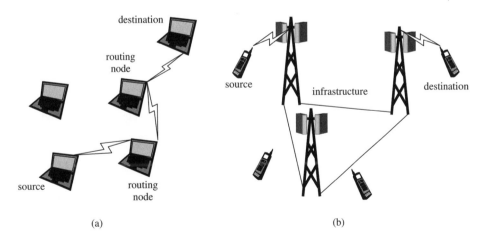

Figure 2.1   (a) Ad hoc and (b) cellular network

Security in Wireless Ad Hoc and Sensor Networks   Erdal Çayırcı and Chunming Rong
© 2009 John Wiley & Sons, Ltd

Table 2.1 Ad hoc and cellular networking

	Ad hoc networking	Cellular networking
Infrastructure	There is no infrastructure.	There is a fixed infrastructure.
Topology	The backbone nodes may be mobile. Topology may change, often due to mobility and/or node failures.	The nodes in the infrastructure are fixed. Terminals can be mobile. However, the topology of the infrastructure seldom changes.
Nodes	The terminal nodes used by the users can also relay the traffic of other nodes.	The nodes in the infrastructure convey data between the source and destination. Their usage as terminal stations or host computers is not mundane. The terminal nodes do not relay traffic from others.
Links	The links are mostly wireless. An end-to-end connection can be made through multiple wireless links, i.e. hops.	The terminal nodes access the infrastructure via a wireless link. The links in the infrastructure can be wireless or nonwireless.

2.1 Ad Hoc Networks and Applications

In wireless ad hoc networks, nodes are generally tetherless, which provides them with the freedom to move. They may act as both host and router, and can relay other nodes' traffic. When there are self-forming, self-configuring and self-healing algorithms for these types of node, a very flexible and rapidly deployable networking architecture that can fulfill the requirements of various applications can be developed.

Tactical communications and military networks are the most obvious applications for ad hoc networks. We cover them in a separate subsection. Apart from tactical communications, there are many other application areas that require rapid deployment and tetherless communications.

2.1.1 Application Examples

The most obvious application areas for ad hoc networks include, but are not limited to, the following:

- **Temporary network deployment:** ad hoc networks can be deployed when it is not viable or cost effective to construct an infrastructure. For example, they can be used as a temporary solution in conferences, underdeveloped or sparsely populated areas and on terrain on which it is too difficult to install an infrastructure.
- **Disaster relief operations:** the rapid deployment capability of ad hoc networks makes them an eminent technology to use for the management of relief operations after large-scale disasters such as earthquakes, tsunamis and floods.
- **Smart buildings:** a large number of sensors and actuators can be deployed without installing any infrastructure to create smart surroundings and a sentient computing environment.

- **Cooperative objects (COs):** COs are entities that are composed of sensors, actuators and COs capable of communicating and interacting with each other and with the environment in a smart and autonomous way to achieve a specific goal. Note that this is a recursive definition. A CO can be composed of other COs. COs are often mobile and sentient entities that react to real-time sensed data coming from a large number of sensors embedded in the environment as well as requests coming from other COs in the vicinity.
- **Health care:** systems to monitor the health conditions and whereabouts of patients and elderly people form another obvious application area for ad hoc networks.

2.1.2 Challenges

There are many more application areas for ad hoc networks. These applications can be realized by tackling challenges specific to wireless ad hoc networking. Some of these challenges are briefly explained below:

2.1.2.1 The Wireless Medium

In ad hoc networks at least some of the communication links are established through the wireless medium. The wireless medium is differentiated from other media mainly by the following:

- the wireless medium is more error prone; for example, the bit error rates (BERs) in the wireless medium can be 10^7 times higher than fiber optic;
- the errors are burstier than guided media;
- the capacity of the wireless medium is limited; when a guided medium is used, it is possible to increase the capacity by laying new lines but in the wireless medium, the spectrum is limited and cannot be extended.

2.1.2.2 Interference, Hidden Terminals and Exposed Terminals

Unconstrained transmission in broadcast media may lead to the time overlap of two or more packet receptions, called *collision* or *interference*. This is also possible in guided media, where the collisions can be detected. However, in the wireless medium the hidden terminal phenomenon hinders the detection of collisions, i.e. the transmission of a terminal can be interfered with by another terminal that cannot be detected. A terminal that can interfere but cannot be detected is called a *hidden terminal*. For example, the transmissions by Sources *a* and *b* interfere with each other at the destination in Figure 2.2. However, neither of the source nodes can sense the transmission by the other because they are hidden from each other, i.e. they are out of range of each other.

Another phenomenon that has an impact on the efficiency of ad hoc protocols, especially for medium access control, is called the *exposed terminal*. Source *a* may not start its transmission to Destination *c* in order to avoid colliding with the transmission of Source *b* to Destination *d*, although in this case neither source interferes with the other's transmission at either destination. Here, Source *a* is an exposed terminal. This is depicted in Figure 2.3.

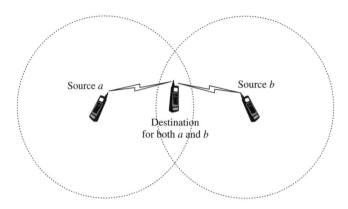

Figure 2.2   Hidden terminal

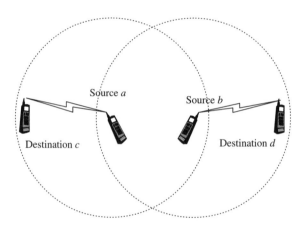

Figure 2.3   Exposed terminal

Hidden and exposed terminals create challenges specific to wireless ad hoc networks. The schemes that tackle these challenges are covered in Chapter 4.

2.1.2.3 Mobility, Node Failures, Self-forming, Self-configuration, Topology Maintenance, Routing and Self-healing

The challenges introduced by the wireless medium are aggravated by the mobility of nodes, which act as both terminals and routers. If the mobile nodes were only terminals, the location management techniques in cellular networks would suffice. However, in ad hoc networks any node can be both a terminal and a router. Therefore, when nodes change their location or fail, the topology of the network that they form changes. Most ad hoc networks are self-forming, self-configuring and self-healing, which means they can autonomously form a network and adapt to the changes in the network. The efficiency in these self-forming and self-healing schemes is closely related to the availability, the level of detail and the accuracy of the topology data at nodes, which constitute the level of network awareness. The network awareness is provided by topology maintenance at each node, and that is not free.

There is a tradeoff between the topology maintenance cost and the efficiency of self-forming and self-healing algorithms. As the resolution and the accuracy of the topology data increase, more efficient self-forming and self-healing algorithms can be developed. However, this also indicates an increase in the topology maintenance cost, i.e. the number of data packets transferred for topology maintenance, which is also dependent on the frequency of topology changes. The topology maintenance process can be classified according to the following criteria:

- traffic generated for monitoring purposes: active or passive;
- monitoring frequency: on demand (event-driven) or continuous (time-driven);
- replication of information: centralized or distributed.

We examine the challenges related to topology maintenance in more detail in Chapter 5.

2.1.2.4 Node Localization and Time Synchronization

In a network where there is no fixed infrastructure, node localization and time synchronization become more challenging. Both of these topics can be very important for security and networking protocols in many applications. Therefore, we dedicate separate sections to them in Chapter 7.

2.1.2.5 End-to-end Reliability and Congestion Control

Since topology changes are always imminent in ad hoc networks and the wireless medium is error prone, the end-to-end connection-oriented transmission control protocol (TCP), which is based on the assumption that the packet losses during transfer are mostly due to congestion, does not fit well with ad hoc networks.

2.2 Sensor and Actuator Networks

A wireless sensor and actuator network (WSAN) is an ad hoc network deployed either inside the phenomenon to be observed or very close to it. Unlike some existing sensing techniques, the position of sensor network nodes need not be engineered or predetermined. This allows random deployment on inaccessible terrain. On the other hand, this also means that sensor network protocols and algorithms must possess self-organizing capabilities.

Another unique feature of sensor networks is the cooperative effort of sensor nodes. Sensor network nodes are fitted with an on-board processor. Instead of sending the raw data to the nodes responsible for the fusion, sensor network nodes use their processing abilities to carry out simple computations locally and transmit only the required and partially processed data.

In a WSAN hundreds to several thousands of sensor nodes (snodes) are densely deployed throughout the sensor field. The distance between two neighboring snodes is often limited to several meters. The node deployment is usually done randomly by scattering nodes in the sensor field. In some applications, actuators (anodes) that control various devices can also be positioned within the sensor network. A collector node (cnode), which is often more capable than the other nodes in the field, is also located either inside or close to the sensor field, as shown in Figure 2.4. Cnodes, usually called *sinks* or *base stations*, are responsible for collecting the sensed data from snodes and then serving the collected data to users. They are also responsible for starting task disseminations in many applications. The sensed data by snodes is conveyed through the sensor network by multiple hops in an ad hoc manner, and gathered in cnodes that can be perceived as the interface between sensor networks and users. Multiple sensor networks can be integrated into a larger network through the Internet or direct links between either cnodes or gateways.

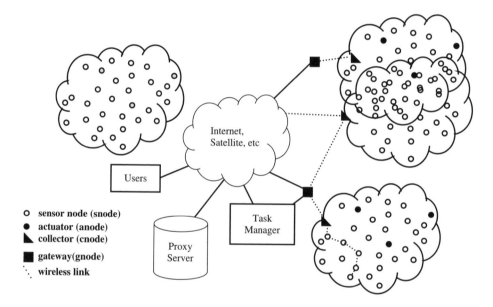

Figure 2.4   Wireless sensor and actuator networks

2.2.1 Application Examples

WSANs have a wide range of potential applications, including security and surveillance, control, actuation and maintenance of complex systems and fine-grain monitoring of indoor and outdoor environments. Some examples of these applications are explained below:

- **Military applications:** wireless sensor networks may be an integral part of military command, control, communications, computers, intelligence, surveillance, reconnaissance and targeting (C4ISRT) systems. The rapid deployment, self-organization and fault-tolerance characteristics of sensor networks make them a very promising sensing technique for military C4ISRT. Since sensor networks are based on the dense deployment of disposable and low-cost sensor nodes, destruction of some nodes by hostile action does not affect a military operation as much as the destruction of a traditional sensor. Some of the military applications are friendly force tracking, battlefield surveillance, reconnaissance, targeting, battle damage assessment and chemical biological radiological and nuclear (CBRN) detection.

- **Environmental applications:** some environmental applications of sensor networks include tracking the movements of species, i.e. habitat monitoring, monitoring environmental conditions that affect crops and livestock, irrigation, macro instruments for large-scale Earth monitoring and planetary exploration and chemical/biological detection.
- **Commercial applications:** there are many potential and emerging commercial WSAN applications such as inventory management, product quality monitoring, smart offices, patient and elderly monitoring, material fatigue monitoring and environmental control in office buildings. There are also some more futuristic WSAN applications such as medical implant communication services, where numerous sensors and actuators are implanted in the human body for various purposes such as continuous monitoring, artificial immune system creation and paralyzed muscle stimulation.

2.2.2 Challenges

WSANs differ from conventional network systems in many respects. They usually involve a large number of spatially distributed, energy-constrained, self-configuring and self-aware nodes. Furthermore, they tend to be autonomous and require a high degree of cooperation and adaptation to perform the desired coordinated tasks and networking functionalities. As such, they bring about new challenges in addition to those introduced by conventional wireless ad hoc networks (Akyildiz *et al.*, 2002).

2.2.2.1 Topology Changes

Snodes may be statically deployed in some WSANs. However, device failure is a common event due to energy depletion or destruction. It is also possible to have sensor networks with highly mobile nodes. Besides, snodes and the network experience vary in task dynamics, and they may be targets for deliberate jamming. Therefore, sensor network topologies may be prone to more frequent changes than conventional ad hoc networks.

2.2.2.2 Fault Tolerance

Sensor networks should be able to sustain their functionalities without any interruption due to snode failures. Protocols and algorithms may be designed to address the level of fault tolerance required by the sensor network applications. The requirements of applications usually differ from each other. For example, the fault-tolerance requirements of a tactical sensor network may be considered much higher than those for a home application because snodes are prone to higher failure rates in tactical sensor networks and the impact of sensor network failure in a tactical field can be much more important than the impact of a home sensor network failure.

Differences in the requirements of various sensor network applications can be observed for almost every factor influencing the design of sensor networks. Moreover, tradeoffs are generally required among these factors because there are stringent constraints related to them. Therefore, a 'one size fits all' generic design is not possible for many tasks in sensor networks. Generally, different schemes are needed to fulfill the requirements of different applications.

2.2.2.3 Scalability

The number of snodes deployed in a sensor field may reach millions in some applications. Moreover, the node density may be as high as 20 nodes/m^3 in some applications. All schemes developed for sensor networks have to be scalable enough to cope with the node densities and numbers, which are higher than all other types of network in terms of orders of magnitude.

2.2.2.4 Sensor Node Hardware

An snode is made up of four basic components: sensing units, a processing unit, a transceiver unit and a power unit. They may also have application-dependent additional components such as a location-finding system, a power generator and/or a mobilizer. Sensing units are usually composed of two subunits: sensors and analog-to-digital converters (ADCs). The analog signals produced by the sensors based on the observed phenomenon or stimuli are converted to digital signals by the ADC and then fed into the processing unit. Note that an snode may be attached to more than one sensor. For example, a temperature and a humidity sensor may be attached to the same snode. The processing unit, which is generally associated with a small storage unit, manages the procedures that make an snode collaborate with the other nodes to carry out the assigned sensing tasks. A transceiver unit connects the node to the network. One of the most important components of an snode is the power unit. Power units may be supported with an energy-scavenging tool such as solar cells. There are also other subunits, which are application dependent. Most sensor network routing techniques and sensing tasks require knowledge of the location with a high degree of accuracy. Thus, it is common for an snode to have a location-finding system. A mobilizer may sometimes be needed to move snodes when this is required to carry out the assigned tasks. All of these subunits may need to fit into a matchbox-sized module. The required size may even be smaller than a cubic centimeter in some applications.

2.2.2.5 Production Costs

As stated in Section 2.2.2.3, a sensor network may contain millions of snodes. Therefore, the cost of snodes has to be low in order for such network to be feasible.

2.2.2.6 Environment

Snodes are densely deployed either very close to or directly inside the phenomenon to be observed. Therefore, they usually work unattended in remote geographic areas, often in extremely harsh environments. They work under high pressure at the bottom of oceans, in difficult environments such as debris or battlefields, under extreme temperatures, such as in the nozzle of an aircraft engine or in Arctic regions, and in extremely noisy environments such as under intentional jamming.

2.2.2.7 Power Consumption

The wireless sensor network nodes can only be equipped with a limited energy source. In some application scenarios, replenishment of power resources might be impossible. Therefore, sensor node lifetime shows a strong dependence on battery lifetime. Hence, power conservation and power management take on additional importance. In other mobile and ad hoc networks, power consumption has been an important design factor, but not the primary consideration, simply because power resources can be replaced by users. In sensor networks, power efficiency is an important performance metric, directly influencing network lifetime.

Power consumption in sensor networks can be divided into three domains: *sensing*, *communication* and *data processing*. Sensing power varies with the nature of applications. Data communication is a major reason for energy consumption. This involves both data transmission and reception. It can be shown that for short-range communication with low radiation power, transmission and reception energy costs are nearly the same. Another important consideration related to data communications concerns the path loss exponent, λ. Due to the low-lying antennae, λ is close to 4 in sensor networks. Therefore, routes that have more hops with shorter distances can be more power efficient than routes that have fewer hops with longer distances.

Energy expenditure on data processing is much lower than on data communication. The example described in Pottie and Kaiser (2000) effectively illustrates this disparity. Assuming Rayleigh fading and fourth power distance loss, the energy cost of transmitting 1 Kb a distance of 100 m is approximately the same as that for executing 3 million instructions with a 100 million instructions per second (MIPS)/W processor.

2.3 Mesh Networks

In wireless mesh networks (WMNs), each node may also be both a router and a host. There are two basic types of mesh networking nodes, namely *mesh routers* and *mesh clients*. Apart from these, other types of nodes, such as personal computers (PCs), personal digital assistants (PDAs), televisions (TVs) or audio sets and video cameras, can also connect to a WMN via a mesh router. A mesh router usually has at least two radios and various network interfaces such as IEEE 802.3, IEEE 802.11, IEEE 802.15 and IEEE 802.16, and therefore it can access other types of network such as the Internet, cellular, LANs, wireless LANs or other ad hoc networks. The mesh router usually uses one of the radios to communicate with mesh clients while using the other radio to route data packages to other mesh routers.

We can further categorize mesh routers as *access*, *backbone* or *gateway* mesh routers. Mesh clients access a mesh network through an access mesh router, and a mesh backbone is connected to an external network such as the Internet through a gateway mesh router. A single mesh router can also carry out all these functionalities. For example, the mesh router in the access mesh in Figure 2.5 is an access, backbone and gateway mesh router.

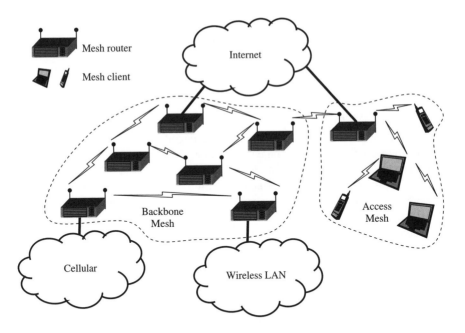

Figure 2.5 Wireless mesh networks

There are two types of mesh networks: *backbone mesh* (BM) networks and *access mesh* (AM) networks. A BM can be formed by using mesh routers, as depicted in Figure 2.5. Alternatively, an AM can be created by mesh routers and mesh clients. Moreover, an AM can be connected to a BM. It is also possible to integrate the Internet or other types of terrestrial or satellite networks into a WMN. Compared to conventional ad hoc networks and WSANs, a WMN can best be distinguished by the following four points:

- Mesh routers are more capable and have fewer energy constraints than the nodes in WSANs and conventional ad hoc networks, and can convey broadband traffic.
- Although an AM or a BM can be formed in a local area, a BM can also be used to provide connectivity in wider areas.
- A BM can be used to integrate various types of network, such as the Internet and a WLAN, into each other through a wireless backbone.
- In ad hoc networks and WSANs it is possible to cluster nodes and provide a hierarchical topology where the cluster heads use one radio to communicate with the nodes in the cluster and another radio to provide the connectivity among the cluster heads. This is the main design approach in mesh networks where mesh routers have at least two radios – one for the mesh clients and the other for forming the backbone with the other mesh routers.

2.3.1 Application Examples

Almost every ad hoc network application scenario can also be implemented as a WMN. However, the main difference between a WMN and other ad hoc networks is based on the capability to transfer broadband data over long distances via a self-forming and self-healing architecture. Therefore, mesh networks are an eminent technology for wireless broadband backbone networks in metropolitan and wide areas. The features and characteristics of WMNs enable the design of other applications such as the following (Akyildiz *et al.*, 2005):

- broadband home networking;
- community and neighborhood networking;
- enterprise networking;
- transportation systems;
- building automation and control networks.

2.3.2 Challenges

All the challenges defined for ad hoc networks are also applicable to WMNs. In addition to these, more capable mesh routers that can transmit broadband traffic over longer distances are required for a WMN. A BM integrates various types of wireless and nonwireless networks. This introduces some new challenges, especially in the transport layer. Guaranteeing the quality of service in this heterogeneous environment is also an important issue. Finally, WMNs present new reliability and security challenges because many mesh routers in a WMN are in the custody of individuals.

2.4 Tactical Communications and Networks

Many mobile tactical networks are actually wireless ad hoc networks. Moreover, most obvious WSAN applications are in the tactical field. Therefore, we dedicate this section to tactical communications. However, please note that our goal is not to explain a specific tactical communications system owned by a nation or an organization. Instead, we very briefly introduce the design principles and challenges for tactical communications in this section.

The essence of designing a good tactical communication system is to enhance survivability and rapid deployment capability. Since this goal must often be achieved in a very harsh and hostile environment, tactical communication systems are one of the most challenging application areas of communications. Other important characteristics of tactical communications are (Cayirci and Ersoy, 2002; Onel *et al.*, 2004):

- **Various mobility patterns:** while some subscribers move at supersonic speeds, others may be fixed.
- **Wide range of terminal types:** a wide range of equipment such as sensors (video camera, radar, sonar, thermal camera, etc.), single channel radios and computers may be the terminals of a military communication network.
- **Variable communication distances:** communication distances range from several meters up to thousands of kilometers.
- **Variable communication medium characteristics:** various types of media such as wires, optical fibers, the air and the sea may be used.
- **Rapidly changing communication locations:** the regions to be covered by extensive communication networks may need to be emptied and the same networks installed in different regions within days during a military operation.

- **Hostile and noisy environments:** the communication facilities of the opposing side are high-priority targets in a battlefield. In addition, thousands of exploding bombs, vehicles and intentional jamming cause noise.
- **Bursty traffic:** the communications traffic is often both temporally and spatially correlated. Long periods of radio silence can, all of a sudden, be broken by extremely heavy reporting and communications requirements in certain regions while other regions maintain silence.
- **Various types of application:** the military communication networks host various types of applications that need to meet different end-to-end quality of service requirements.
- **Various security constraints:** unclassified data together with top-secret data flow through the same communication channels.

In order to satisfy these requirements, intensive research projects have been conducted, such as Defense Information System (DISN), Post-2000 Tactical Communications (TACOMS) and Global Mobile Communications. Figure 2.6 illustrates the architecture of the state-of-the-art tactical communications systems derived from DISN and TACOMS efforts. This architecture has four subsystems: the *local area subsystem* (LAS), the *wide area subsystem* (WAS), the *mobile subsystem* (MS) and the *system management and control subsystem* (SMCS). A security system is also integrated into the architecture. WAS interconnects other subsystems as a wide area backbone. It is generally a predeployed high-capacity network designed and managed better than the other subsystems. It may be deployed and used in peacetime. The LAS can

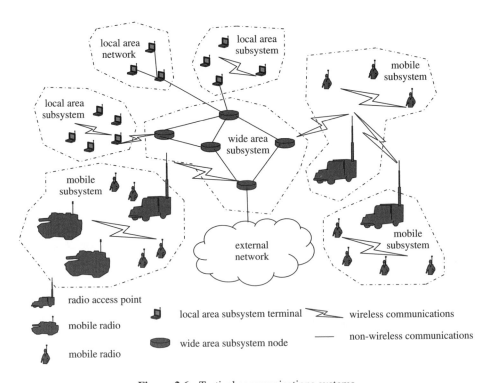

Figure 2.6 Tactical communications systems

be considered a nomadic *local area network* that can access the WAS or available commercial networks. Headquarters and similar organizations that sustain their presence in restricted areas are provided with local area networking support by an LAS. Mobile users in a battlefield access a tactical communication system through an MS. An MS may operate as an independent communication network or may be a part of the overall tactical communication system by accessing the WAS. The SMCS, a subsystem integrated into the architecture, provides the network administrators with system management functions.

Among these subsystems an MS is the most important candidate for employing WASM technologies. An MS has two important technology components, namely *mobile radios* (MRs) and *radio access points* (RAPs). A user of an MS accesses the integrated services provided by a tactical communications system through an MR, which is a terminal station most of the time. However, an MR in an MS can also relay other traffic, just like a wireless ad hoc network node. An RAP conveys the multimedia traffic among MRs, and between a WAS and an MS. In many respects, they carry out functions similar to those of a mesh router.

An MS has a rapidly deployable mobile infrastructure and uses both cellular and ad hoc techniques. In Figure 2.7, the MS architecture for tactical communication systems is illustrated. In this MS architecture there are four tiers:

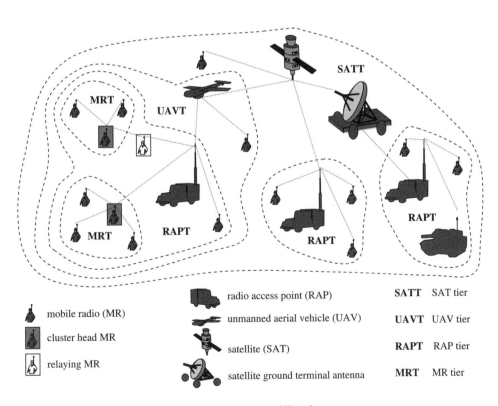

mobile radio (MR)	radio access point (RAP) **SATT** SAT tier
cluster head MR	unmanned aerial vehicle (UAV) **UAVT** UAV tier
relaying MR	satellite (SAT) **RAPT** RAP tier
	satellite ground terminal antenna **MRT** MR tier

Figure 2.7   Multitier mobile subsystem

- **Mobile radio tier (MRT)**: an MRT cell is a cluster of MRs, which cannot access any RAP. In each MRT cell (i.e. an MR cluster), one of the MRs becomes the head of the cell (i.e. cluster). If there is an MR that can access a RAP and can be accessed by the MRT cell head, then the MRT cell is connected to the RAP via the MR, which relays the traffic between the MRT cell and the RAP.
- **Radio access point tier (RAPT):** an RAP is the mobile base station of an MS. They create RAPT cells. An RAPT cell may also construct an underlay cluster for the other MRT cells when some MRT cells are connected to them by relaying MRs.
- **Unmanned aerial vehicle tier (UAVT):** this is the first level overlay tier of an MS. The UAVT cells cover the areas that are not covered by the lower tiers. The lower tier cells can also use UAVs to access the WAS.
- **Satellite tier (SATT):** this is the topmost overlay tier. A SATT cell is created by a satellite and may include a number of lower tier cells. Satellites are used by lower tier cells to access the WAS and to communicate with other cells.

The key system requirements for such a military communications system include the following:

- multimedia communications;
- multitier networking;
- mobile networking;
- mobile and rapidly deployable infrastructure;
- survivable infrastructure;
- tailorable infrastructure;
- multifunctional infrastructure;
- modular infrastructure;
- flexible infrastructure;
- both terrestrial and nonterrestrial networking;
- horizontal and vertical communications ability;
- high circuit quality and wide bandwidth;
- secure networking;
- real-time and batch networking;
- ability to operate in all weather and terrain conditions.

2.4.1 Blue Force Tracking System

A Blue Force Tracking (BFT) system is made up of three main components: a computer, a wireless transceiver (typically a satellite transceiver) and a Global Positioning System (GPS) receiver. These three components are integrated and put into a vehicle. The GPS locates the vehicle and the computer displays its location on the geographical information system available in the computer along with other vehicles in the vicinity. A BFT also provides interfaces for text messaging and reporting of the information available in the computer. This capability is also used to report the locations of enemy forces and other battlefield conditions.

Such a system has many uses, among which the situational awareness provided to decision makers at all levels is the most important. A commander can follow the whereabouts of his/her units by BFT. Artillery and aircraft can have a better view of the locations of friendly forces and can avoid engaging with friendly forces by mistake. Smart and more efficient logistics systems can be developed based on BFT. We can expect BFT nodes to be deployed more widely with all troops, vehicles, critical equipment, combat systems and ammunition to track both their locations and conditions. Presumably they will be a combination of terrestrial ad hoc networks connected to each other through larger gateway nodes that can access a satellite network.

Almost every modern army has plans to deploy BFT to all units, and security is a big issue in BFT. Every challenge that we have outlined for ad hoc and sensor network security is also valid for BFT. Indeed this is the one application where almost everything available for securing ad hoc and sensor networks is needed. Therefore, this short section emphasizes the importance of wireless ad hoc and sensor network security for BFT.

2.5 Factors Influencing the Design of Wireless Ad Hoc, Sensor and Mesh Networks

In every section of this chapter we elaborate the relevant challenges, which actually equate to the factors influencing the design of WASM protocols and algorithms (Table 2.2). In this section we discuss these factors in a more holistic manner.

Table 2.2 Factors influencing wireless ad hoc, sensor and mesh network design

Factor	Ad hoc	Mesh	Sensor and actuator
Wireless medium	ISM	ISM	ISM, acoustic, low-lying antenna
Networking regime	random one-to-one	random one-to-one, gateway nodes	one-to-many, many-to-one, many-to-many
Traffic	random, multimedia	random, multimedia	temporally and spatially correlated data
QoS requirements	bandwidth, delay, jitter, reliability	bandwidth, delay, jitter, reliability	power consumption, delay, reliability
Mobility	mobile	typically fixed	generally fixed, network mobility
Fault tolerance	typically no critical point of failure	critical points of failure	critical points of failure, high fault-tolerance requirements
Operating environment	typical day-to-day environment	typical day-to-day environment	hostile and harsh, often unreachable
Power efficiency	not very critical	not critical	critical
Scalability	order of hundreds	order of tens	order of thousands
Hardware constraints	laptops, PDAs	no constraint	tiny, low processing and memory capacity
Production cost	no hard constraints	no hard constraints	must be cost effective

2.5.1 Wireless Medium

The only viable medium for most of the ad hoc applications is wireless, i.e. radio, infrared or optical, which has limitations and challenges, as already outlined. To fulfill requirements such as ad hoc deployment and global operation, the chosen transmission medium must be available worldwide and license-free. Therefore, one option for radio links is the use of industrial, scientific and medical (ISM) bands, which offer license-free communication in most countries. The International Table of Frequency Allocations specifies the frequency bands shown in Table 2.3 as the bands that may be made available for ISM applications. Some of these frequency bands are already being used for communications in cordless phone systems and wireless local area networks (WLANs). Therefore, these bands are crowded with many wireless systems and applications. However, they still have advantages such as the globally available free and huge spectrum allocation. They are not bound to a particular standard, thereby giving more freedom for implementation. On the other hand, there are various rules and constraints, like power limitations and harmful interference from existing applications.

Table 2.3 ISM
frequency bands

6765–6795 kHz
13 553–13 567 kHz
26 957–27 283 kHz
40.66–40.70 MHz
433.05–434.79 MHz
902–928 MHz
2400–2500 MHz
5725–5875 MHz
24–24.25 GHz
61–61.5 GHz
122–123 GHz
244–246 GHz

Another possible mode of inter-node communication in ad hoc networks is by infrared. Infrared communication is license-free and robust to interference from electrical devices. Infrared-based transceivers are cheaper and easier to build. Many of today's laptops, PDAs and mobile phones offer an infrared data association (IrDA) interface. The main drawback of infrared is the requirement of a line of sight between sender and receiver. This makes infrared a reluctant choice of transmission medium in the ad hoc network scenario.

An interesting development is the Smart Dust mote, which is an autonomous sensing, computing and communication system that uses the optical medium for transmission. Two transmission schemes, passive transmission using a corner-cube retro-reflector (CCR) and active communication using a laser diode and steerable mirrors, are examined for Smart Dust. In the former, the mote does not require an onboard light source. A configuration of three mirrors (CCR) is used to communicate a digital high or low. The latter uses an onboard laser diode and an active-steered laser communication system to send a tightly collimated light beam toward the intended receiver.

The unusual application requirements of ad hoc networks make the choice of transmission medium more challenging. For instance, marine applications may require the use of the aqueous transmission medium. Oceans offer a very different propagation environment which is not appropriate for RF communications. Acoustic channels lower than 100 kHz present the only carrier option for the time being. Acoustic signals travel 10^5 times slower than RF, and introduce very high propagation latency, i.e. 67 ms for 100 m. The extremely high latency impacts on the performance of MAC, network and transport layer protocols.

Sensor networks differ from ad hoc and mesh networks mainly because of the stringent energy constraints, low capacity and small size of tiny sensor nodes. In sensor networks, energy minimization assumes significant importance, over and above the decay, scattering, shadowing, reflection, diffraction, multipath and fading effects. In general, the minimum output power required to transmit a signal over a distance d is proportional to d^n, where $2 \le n < 4$. The exponent n is closer to four for low-lying antennae and near-ground channels, as is typical in sensor network communication. Therefore, routes that have more hops with shorter hop distances can be more power efficient than those that have fewer hops with longer hop distances.

The wireless medium makes ad hoc networks vulnerable because it is easy to tap. Transmission power constraints and shorter ranges bring up both disadvantages and advantages. Jamming is easier against an adversary with low power transmissions. On the other hand, most of these systems are more resilient to interference, and a node needs to be close to tap low transmission power systems.

2.5.2 Networking Regime

One of the key distinctions between WMNs and conventional ad hoc networks is that the traffic is typically to or from a gateway node connected to the Internet in WMNs but between a random pair of nodes in conventional ad hoc networks. This is similar to WSANs, where the traffic is typically to or from the data collection node (cnode), i.e. sink. However, WSAN traffic is not the same as WMN traffic. While WMN traffic generally has a one-to-one nature, traffic is either one-to-many or many-to-one in typical sensor networks, where a cnode sends a task to snodes or snodes report their results to a cnode. When actuators are included, this relation can become many-to-many, where multiple snodes report their measurements to multiple anodes. Similarly, there is a hierarchical nature in tactical communications where the traffic is typically to or from the higher echelons. Note that this is not always the case, and there can be WSAN or WMN applications where the source and destination are random nodes within the network.

The existence of a point of gravity or a key node in a WSAN or WMN makes them more vulnerable in a hostile environment. It is possible to discover a mesh router or cnode by analyzing the traffic, and to monitor or to prevent all data traffic in these key nodes.

In every kind of ad hoc network, i.e. conventional, mesh or sensor, nodes rely on each other to deliver a package. This multihop self-organizing nature also introduces additional weaknesses, making them vulnerable to attack. When a malicious node convinces other nodes to be a relaying node for it, it can receive their packets and may not relay them. These attacks and challenges will be examined in more detail in the following chapters.

2.5.3 Nature of Traffic

Apart from the networking regime, ad hoc, sensor and mesh networks also have other distinctive characteristics related to the traffic. In ad hoc and mesh networks, the data generation rate is generally random and depends on the habits and specifics of the users and applications. On the other hand, in sensor and tactical networks, data traffic is usually both temporally and spatially correlated. Sensor coverage areas generally overlap, and therefore, when an event occurs, it triggers multiple sensors in the same region. Similarly, nodes tend to keep silent until contacting the enemy, when everyone in the contact region starts reporting. Temporal and spatial correlation indicates both overutilization for some areas and time periods and underutilization for other areas and time periods. This introduces additional challenges for the design of communication protocols and algorithms, including security schemes. When data traffic is correlated, defending against traffic analysis attacks becomes more challenging.

2.5.4 Quality of Service Requirements

The quality of service (QoS) for a data flow is generally characterized by three parameters: bandwidth, delay and reliability. There is another parameter, related to variations in the delay, called *jitter*.

First, the nature of the applications in conventional ad hoc networks and mesh networks is not very different from those in other local or wide area networks. Audio, video and data traffic are conveyed by them, and the QoS requirements for this traffic do not change in a wireless environment. However, guaranteeing the required QoS levels is more challenging due to the limited spectrum, high bit error rates (BERs) and temporal BER variations in the wireless medium, in addition to the mobility and self-organization issues introduced by ad hoc networking.

In sensor networks, challenges are exacerbated by the stringent power constraints. In many sensor network applications power is the primary concern. Of course this depends on the application. When sensor and actuator networks are used for real-time applications, delay is also an important constraint, generally conflicting with the power constraints. Also, bandwidth requirements for sensor networks may be high. For example, long-time acoustic measurements require large bandwidth to be sent back loss-free. There are also some sensor network applications where delay and bandwidth issues are the most important challenges. An underwater acoustic network is an example of this. Propagation latency is very high (67 ms for 100 m on average) and capacity is very limited (5–30 kbps) for the acoustic underwater medium.

2.5.5 Mobility

Mobility is one of the key features for many ad hoc networks. The nodes may be randomly and independently moving around in many mesh and ad hoc networking applications. This is again different for many sensor networks, where nodes generally move together. Although there may be small changes in direction and speed for each node, they tend to keep their relative positions in the network because nodes are mobile due to a common force. For example, maritime surveillance networks drift with the current if they are not anchored.

2.5.6 Fault Tolerance

Although self-forming and self-healing make ad hoc networks more fault tolerant than fixed networks, mesh routers in mesh networks and the collecting nodes in sensor networks present critical points for failure. The data are relayed through these nodes to the external systems and the network becomes unconnected if they do not exist. This is very important, especially in sensor networks, because the data collected by sensors do not serve any purpose if they do not reach users and the data in the sensor nodes are only accessible through the collecting nodes. Therefore, they may become an important target for denial-of-service security attacks, and fault-tolerance schemes should take this into account.

2.5.7 Operating Environment

Operating environments for conventional ad hoc and mesh networks usually do not impose special handling compared to fixed networks, except that they may be deployed in wide areas with less infrastructure. This should only indicate additional administrative concerns for the service providers. However, sensor networks are designed to run unattended in harsh and inaccessible areas. This introduces additional challenges for fault-tolerance schemes. Moreover, sensor networks may be in an adversarial environment beyond enemy lines. In such cases, they are physically vulnerable and easier to tamper with.

2.5.8 Power Efficiency Requirements

Conventional ad hoc and mesh networks do not have very stringent power constraints. However, power consumption is one of the most important factors influencing sensor network protocol design, which also requires special handling of security issues. Security schemes for sensor networks need to be low cost, both computationally and in the networking requirements.

2.5.9 Scalability

In terms of scalability, sensor networks are again different from the other two types of ad hoc network. Every sensor network scheme needs to be highly scalable, as explained in Section 2.2. This also has an impact on security protocols. For example, the scalability requirement together with power constraints hampers the applicability of post-deployment key distribution schemes for many sensor network applications. Generally, in this kind of application, keys are installed before a node is deployed.

2.5.10 Hardware Requirements and Production Cost

Hardware is less restrictive in mesh and ad hoc networks compared to sensor networks, where nodes have limited memory and computational power. Therefore, security schemes that have less memory and computation requirements are preferable for sensor and actuator networks.

2.6 Review Questions

2.1 How can you distinguish an ad hoc network from a cellular network?

2.2 What are the disadvantages and advantages of the wireless medium compared to the fiber optic medium?

2.3 What are the consequences of hidden and exposed terminals?

2.4 What is topology management? How do you categorize topology management schemes?

2.5 Why does TCP in the TCP/IP suite not perform as well in ad hoc networks as in fixed networks? What are the performance metrics in this comparison and which factors influence the performance of TCP?

2.6 What are the impacts of actuators on the networking regime in wireless sensor and actuator networks?

2.7 What are the main differences between access and backbone mesh networks?

2.8 What are the differences between mobile and local area subsystems in tactical communications?

2.9 What do you understand by horizontal and vertical communications?

2.10 What are the main differences between the acoustic and RF media?

3

The Wireless Medium

In this chapter we explain the fundamental concepts related to wireless communications. The material provided is introductory and related to the security issues. For interested readers, further details about the physical layer data communication concepts can be found in Stallings (2000).

3.1 Wireless Channel Fundamentals and Security

Electromagnetic signals can be either analog or digital. The intensity of an analog signal varies and the change in the signal strength is usually smooth and continuous. On the other hand, digital signal strength is constant for a period, after which it may change to another discrete level. Analog and digital signals are depicted in Figure 3.1.

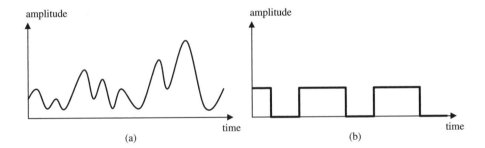

Figure 3.1 (a) Analog and (b) digital signal

When the signal strength is the same as the signal strength a certain time period, T, before, and this repeats every T, the signal is called periodic. Figure 3.2 shows sine and square waves, which are analog and digital periodic signals respectively, where

$$s(t) = s(t - T) \tag{3.1}$$

Security in Wireless Ad Hoc and Sensor Networks   Erdal Çayırcı and Chunming Rong
© 2009 John Wiley & Sons, Ltd

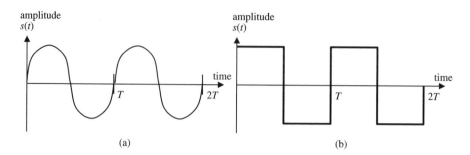

Figure 3.2   (a) Sine and (b) square wave

The *frequency (f)* of a periodic signal is the rate at which it repeats a complete cycle in every second, i.e. $f = 1/T$, and this rate is given in Hertz (cycles per second). For example, a sine wave completes a 2π, i.e. 360°, cycle in every T and starts a new cycle. If the peak *amplitude A, frequency f* and initial *phase ϕ* of a sine wave are given, then the signal strength at time t is

$$s(t) = A \sin(2\pi ft + \phi) \tag{3.2}$$

The sine wave in Figure 3.2(a) has a single frequency f, and Equation (3.2) is enough to represent it as a function of time. However, a typical electromagnetic signal has more than a single frequency component and therefore electromagnetic signal strength also needs to be represented as a function of frequency. For example, the signal in Figure 3.3(d) has three frequency components. The initial phase ϕ is zero for all three components, which means the amplitude of the sine wave is zero initially for all three frequencies. The peak amplitude A for the first frequency f shown in Figure 3.3(a) is *one*. The second frequency is three times higher, i.e. $3f$, and has a peak amplitude three times lower, i.e. $A/3$, compared to the first component. The third frequency is $5f$ with a peak amplitude $A/5$. This signal covers the frequencies between f and $5f$, and has $4f$ *bandwidth*. For example, if f is 2 MHz, then the bandwidth of this signal is 8 MHz.

When we sum the waveforms at frequencies f, $3f$ and $5f$ in Figure 3.3, we have the waveform shown in Figure 3.3(d). It looks like a digital square wave. The more frequency components we add into this signal, the closer it gets to a square wave and the less distortion is observed. Actually, when we have an infinite number of frequencies, it becomes a square wave. In other words, a square wave has an infinite number of frequency components and can be represented by

$$s(t) = \frac{4A}{\pi} \sum_{kodd, k=1}^{\infty} \frac{1}{k} \sin(2\pi kft) \tag{3.3}$$

Note that as the frequency of each additional component increases, its peak amplitude decreases, i.e. the peak amplitude for kf is A/k. Therefore, most of the energy is in the first few frequency components. Moreover, the attenuation is different for each frequency component. Therefore, some of the frequency components are diminished during transmission and the received signal is distorted. The range of frequencies that reach the receiver gives

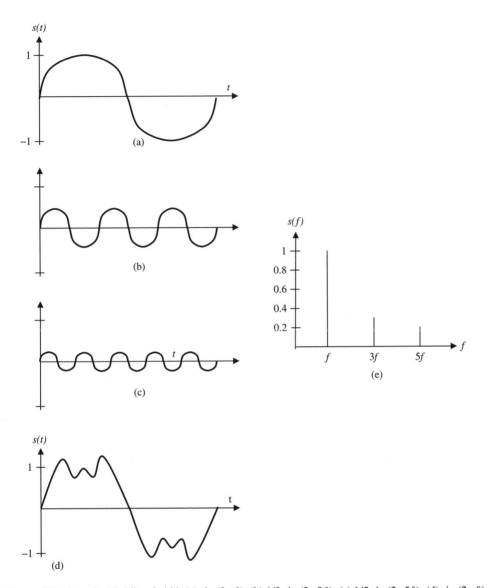

Figure 3.3   Signal with 4f bandwidth (a) sin (2πft); (b) 1/3 sin (2$\pi 3ft$); (c) 1/5 sin (2$\pi 5ft$); (d) sin (2πft) + 1/3 sin (2$\pi 3ft$) + 1/5 sin (2$\pi 5ft$); (e) frequency components

the *absolute bandwidth*. For the interested reader, time and frequency domain concepts are explained in more detail in Stallings (2000).

3.1.1 Capacity

We can conclude that a higher number of frequency components indicates less distortion in the received signal, which creates the ability to transmit and receive at higher bit rates. The frequency component that has the highest energy also has an impact on the capacity. We can say that the *capacity* C of a noise-free channel is limited to its absolute bandwidth B, and can be stated as

$$C = 2B \qquad (3.4)$$

Equation (3.4) gives us the signal rate, which is the number of times we can change the signal in a second. If each signal indicates one of more than two values, we can increase the capacity. For example, when each signal may have one of four different voltage levels, two bits can be transmitted at a time. We examine various ways to represent multiple values by a signal when we explain modulation later in this chapter. When the number of values that each signal can carry is M, the capacity C in bits per second (bps) of the channel for absolute bandwidth B in Hertz is

$$C = 2B \log_2 M \qquad (3.5)$$

Equation (3.5) is called the *Nyquist formulation* and gives the upper capacity limit for the available bandwidth. When the channel is not noise-free, noise introduces additional distortion in the received signal and therefore reduces the capacity. Shannon states the capacity of a channel for the received signal strength S and noise level N as

$$C = B \log_2 \left(1 + {}^S\!/\!_N\right) \qquad (3.6)$$

3.1.2 Electromagnetic Spectrum

Capacity is a function of available bandwidth, which is a range in the electromagnetic spectrum shown in Figure 3.4. As the central frequency increases, the potential bandwidth also increases, which indicates higher potential capacity.

For example, the range of frequencies available for fiber optics is between 10^{14} Hz and 10^{15} Hz. The potential bandwidth within this range is 9×10^{14} Hz. This means a huge potential capacity for fiber optics. However, the availabilities and capabilities of the electromechanical devices that generate and relay signals are limiting factors. Therefore, we can use only a small part of the potential capacity for fiber optics.

At the lower end of the electromagnetic spectrum, audio frequency resides. At first glance, this part of the spectrum may seem unimportant for the ad hoc networking domain. However, it provides the most viable medium for underwater acoustic networks. For underwater communications, radio waves over 50 kHz do not provide the required transmission characteristics. The acoustic bandwidth between 200 Hz and 50 kHz is the most appropriate spectrum interval for this purpose, but it introduces several weaknesses and limitations. First, the potential capacity that can be offered by the available bandwidth in this range is limited to approximately 20 kbps when the communications distance is less than 1 km. In addition to this, the propagation speed in an underwater acoustic channel is about 1500 m/s, which is 10^5 times slower than a radio frequency (RF) signal. An acoustic signal can travel 100 m in about 67 ms, and this low speed makes the propagation latency an important concern for any protocol or algorithm design, especially for acoustic multihop networks. The speed of acoustic signals is also a function of depth and temperature. There are layers where the sea temperature changes by one or two degrees suddenly, and acoustic signals are refracted and lost in these layers because their speed changes based on the temperature. Moreover, absorptive losses underwater are high, which creates a practical limit for frequencies above 100 kHz, and acoustic

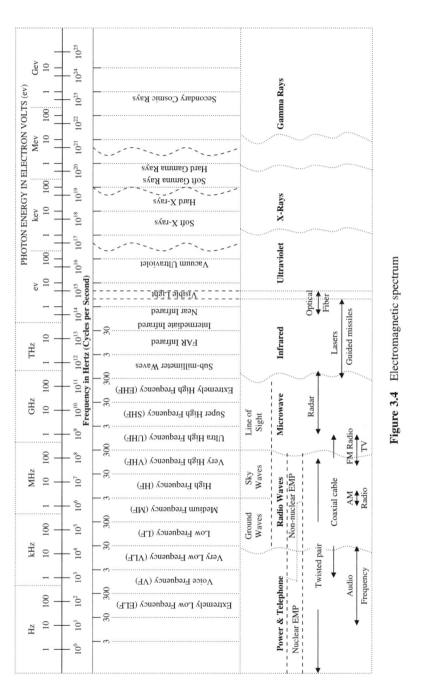

Figure 3.4 Electromagnetic spectrum

signals can be interfered with by additional natural and artificial noise sources such as wind, tides, sea animals and machinery.

Above the acoustic frequency, radio waves start at around 30 kHz. Radio waves are omni-directional, i.e. they are disseminated in all directions, and can penetrate walls and doors. Therefore, they are widely used for indoor and outdoor applications. In the low frequency (LF) and medium frequency (MF) bands, i.e. between 30 kHz and 3 MHz, radio waves follow the ground and are called *ground waves* (Figure 3.5(a)). In the high frequency (HF) and lower half of the very high frequency (VHF) bands, i.e. between 3 MHz and 100 MHz, the signals close to the ground are absorbed. However, they are also refracted and bounced back to Earth by the ionosphere, which is between 100 and 500 km above the Earth. Hence, they can be delivered long distances with the help of the ionosphere and are therefore called *sky waves* (Figure 3.5(b)). Military radios, called VHF radios, communicate by signals in the lower half of the VHF band.

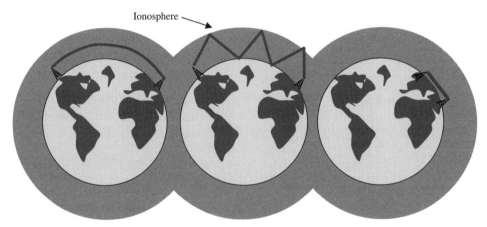

(a) Ground wave (f< 2 MHz) (b) Sky wave (2 MHz <f< 100 MHz) (c) Line of sight (f> 30 MHz)

Figure 3.5 Ground, sky and line-of-sight waves

In the frequency bands above VHF, the waves travel in almost straight lines and direc-tional antennas are often preferred, i.e. the transmissions are generally not omnidirectional. The signals in these bands do not penetrate walls as well as radio waves do. There are also other impairments, which are more effective against microwaves. We examine these in Section 3.1.4.

Since microwaves follow almost straight lines, the curvature of the Earth's surface intro-duces a physical limit for the range of line-of-sight waves (Figure 3.6). This range limit is a function of the antenna height. When the receiving antenna height, h_2, is neglected, the range limit for a microwave can be figured out based on the transmitting antenna height, h_1, by

$$d_1 = 3.57\sqrt{kh_1} \qquad (3.7)$$

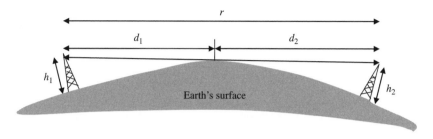

Figure 3.6   The range limit for line-of-sight waves

In Equation (3.7) k is 4/3. The range limit r for a microwave transmission increases when the receiving antenna also has a height.

$$r = 3.57(\sqrt{kh_1} + \sqrt{kh_2}) \tag{3.8}$$

As we discussed in Chapter 2, part of the electromagnetic spectrum is dedicated to the license-free industrial, scientific and medical (ISM) bands. WASM applications generally use the ISM bands. The ISM bands between 433–464 MHz, 902–928 MHz, 2.4–2.5 GHz and 5.725–5.875 GHz are the preferred ISM bands for sensor networks, mesh networks, wireless local area networks (WLANs) and wireless personal area networks (WPANs). Note that these are bands of microwaves and fall within the ultra high frequency and the lower part of the super high frequency designations. Higher frequency microwaves are usually used for satellite communications.

3.1.3 Path Loss and Attenuation

The signal strength of electromagnetic waves is subject to distance-dependent *path loss*. Therefore, the signal power P_r at the receiving antenna is less than the signal power P_t at the transmitting antenna. Although this difference is also related to the antenna gains of the transmitter and receiver, losses in transmission and reception circuitry and other random factors, we can model the difference between P_r and P_t as

$$\frac{P_t}{P_r} = \left(\frac{4\pi d}{\lambda}\right)^{\gamma} = \left(\frac{4\pi f d}{c}\right)^{\gamma} \tag{3.9}$$

where λ is the carrier wavelength, which is the ratio between the speed of light c (3×10^8 m/s) and frequency f, i.e. $\lambda = c/f$, d is the propagation distance between the transmitter and the receiver and γ is the path loss exponent, which is typically two for free space and four for dense urban areas. The path loss exponent can be lower than two when special receiving techniques are used and as high as six in heavy weather conditions and densely obstructed areas. Based on the ratio between the transmitted and received signal power levels, path loss can be expressed in decibels as:

$$L_{dB} = 10 \log \frac{P_t}{P_r} = \gamma \, 10 \log \left(\frac{4\pi f d}{c}\right) \tag{3.10}$$

For free space where the path loss exponent γ is two, Equation (3.10) can be further simplified to

$$L_{dB} = 20 \log \left(\frac{4\pi f d}{c} \right) = 20 \log(f) + 20 \log(d) - 147.56 \, \text{dB} \tag{3.11}$$

As shown in Equation (3.11), a loss in signal strength is not only dependent on the distance; there is also frequency-dependent attenuation. At higher frequencies, attenuation is also higher and electromagnetic signals are more susceptible to absorptive losses by water and water vapour. For example, a signal over a 5 GHz central frequency is much more sensitive to rain, hail and fog than a signal with a central frequency around 1 GHz. This means that higher frequency signals need more transmission power to reach the same range as a lower frequency signal. It also indicates that they fade away in shorter distances after the intended receiver.

Path loss characteristics have implications for security considerations. The higher the frequency, the smaller the controlled space required after the desired communications range, because signals at higher frequencies fade away quicker. This may be considered advantageous for security. On the other hand, the higher frequency also implies higher potential bandwidth. Therefore, when a higher frequency node or carrier is compromised, a greater quantity of data can be transferred to the adversary.

3.1.4 Other Transmission Impairments and Jamming

Apart from attenuation, there are other sources that degrade the signal quality at the receiver. We can categorize these impairments into three broad classes: *noise, distortions by the physical environment* and *Doppler fading*.

3.1.4.1 Noise

Noise is caused by undesired signals interfering with the carrier. There are various forms of noise:

- **White noise:** this is often called thermal noise because it is a function of temperature and due to the thermal agitation of electrons. It cannot be eliminated. Otherwise, when we also eliminate other noise, the capacity of a carrier does not have a theoretical upper limit according to Equation (3.6). Thermal noise is independent of the frequency and is therefore called white noise. For 1 Hz bandwidth, thermal noise is

$$N_0 = kT \tag{3.12}$$

where N_0 is the thermal noise per 1 Hz bandwidth, i.e. watts/hertz, k is Boltzman's constant, i.e. $1.3803 \times 10^{-23} \, \text{J}/°\text{K}$ and T is the temperature in Kelvin. For a bandwidth of W hertz, thermal noise becomes

$$N = kTW \tag{3.13}$$

- **Intermodulation noise:** the signals at two different frequencies f_1 and f_2 might produce a signal at a frequency which is the sum, i.e. $f_1 + f_2$, the difference, i.e. $f_1 - f_2$, or multiples, i.e. $n \times f_k$, of the original frequencies.
- **Crosstalk:** when there is a time overlap of more than two wireless transmissions, and they are received by the same antenna, they interfere with each other. This is synonymous with the crosstalk phenomenon experienced when two telephone lines are unintentionally coupled and communications in one can also be heard in the other.
- **Impulse noise:** irregular, unexpected and very short spikes of noise can be observed in wireless channels. These have various causes and are especially effective on digital communications.

3.1.4.2 Physical Environment

Trees, buildings, etc. can also cause distortions in the signal quality. These are due to the wave propagation phenomena shown in Figure 3.7:

- **Reflection:** when an electromagnetic wave reaches a smooth surface, it is partly absorbed and partly reflected.
- **Diffraction:** if an electromagnetic wave touches a sharp edge, its direction is altered towards the edge.
- **Scattering:** objects like poles and trees in the way of an electromagnetic wave cause multiple copies of the wave to be scattered around.

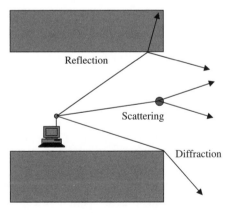

Figure 3.7 Reflection, scattering and diffraction

These phenomena are the reason for another important transmission impairment called *multipath fading*. Apart from the direct signal between the source and the destination, the reflected and scattered copies of the same signal also reach the destination. Since these copies travel farther than the direct copy, they reach the destination with a delay, interfere with the direct copy and degrade the quality of the received signal.

3.1.4.3 Doppler Fading

Doppler fading occurs when the source, destination or both sides of a link move and their relative positions change during communications. As the two ends get closer, the signals travel a shorter and shorter distance, which shifts the signal frequency to higher frequencies. As the two ends move away from each other, a frequency shift occurs towards the lower frequencies. This may cause the receiver to sample the signals at the wrong frequencies.

When any of these impairments result in too low a signal-to-noise ratio (SNR) at the destination, the signal cannot be recovered by the receiver. The well-known denial-of-service attack called *jamming* is based on this phenomenon. To jam a carrier, an intentional noise effective at that frequency is generated such that the SNR for the links in the related frequencies becomes lower than the required level.

3.1.5 Modulation and Demodulation

Digital data need to be converted to waveforms for transmission through a wireless channel. The receiver translates the waveforms back into digital data. These processes are called *modulation* and *demodulation* respectively. Actually, in digital communications the conversion between digital data and analog waveforms is not limited to the sequence *digital–analog–digital*. An analog signal may need to be converted to another analog signal so that it can be carried in a frequency channel other than the original frequency. Moreover, analog signals such as voice may be digitized for more effective storage and transmission. Therefore, analog–analog–analog and analog–digital–analog conversions are also common in wireless communications. Of course, various combinations of these, such as analog–digital–analog–digital–analog, are also possible. For example, voice can be digitized first and then transmitted after modulation. The receiver first demodulates the input waveform and then converts the recovered digital data to the original voice.

For *digital-to-analog* conversion the amplitude, frequency or the phase of a sine wave can be modulated to represent digital data, as depicted in Figure 3.8. Note that the sine wave is an analog periodic signal. In amplitude modulation, a discrete level of amplitude represents 1 while another discrete level represents 0. Similarly, a frequency can indicate 1 and another frequency can indicate 0. Finally, the phase of a sine wave can be changed at the beginning of a symbol duration when the next bit in the input bit stream is different from its predecessor. Alternatively, a discrete phase, e.g. π, at the beginning of a symbol duration may be equal to 1 while another phase, e.g. 2π, may be equal to 0.

The examples in Figure 3.8 represent binary modulation where the amplitude, frequency or phase of the waveform is modulated by using one of two available discrete values during one symbol duration. Symbol duration is the time period in which we sample the waveform once. The inverse of the symbol duration is called the *symbol rate* or *baud rate*. In binary modulation, the data rate or bit rate, which is the number of bits that can be transferred in a second, is equal to the baud rate.

When we can select one of more than two amplitude levels or phases during a symbol duration, the data rate can become higher than the symbol rate. For example, if four amplitude levels are available, we can modulate two bits in each symbol interval, as shown in Figure 3.9(b). Similarly, a sine wave can be shifted through one of four phases, and two bits can again be transferred in each symbol. This later technique is called *quadrature phase*

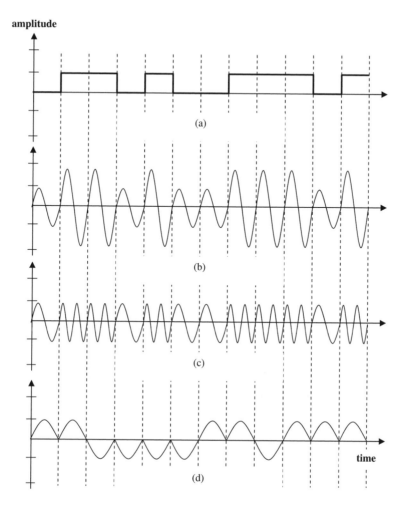

Figure 3.8 Digital-to-analog modulation (a) digital signal; (b) amplitude modulation; (c) frequency modulation; (d) phase modulation

shift keying (QPSK). Moreover, multiple phases and multiple amplitude levels can be used together. For example, four amplitude levels and four phases form 16 combinations in *quadrature amplitude modulation (QAM-16)*, where four bits per symbol can be transferred, and a data rate four times higher than the symbol rate can be achieved. The number of amplitude levels can be further increased until the energy per bit becomes too low to recover the bit at the receiving end. The environments for which the radio is designed should also be considered in determining the ratio of the data rate to the symbol rate. Adversarial environments, where jamming and high interference are expected, may force one to reduce this ratio.

Amplitude and frequency modulation are also commonly used for *analog-to-analog* modulation (Figure 3.10). In amplitude modulation, the amplitude of the wave is modulated

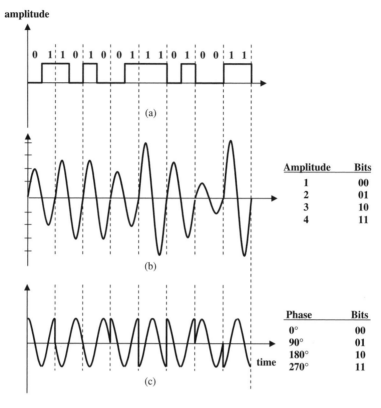

Figure 3.9 Multilevel amplitude and phase modulation (a) digital signal; (b) multilevel amplitude modulation; (c) multilevel phase modulation

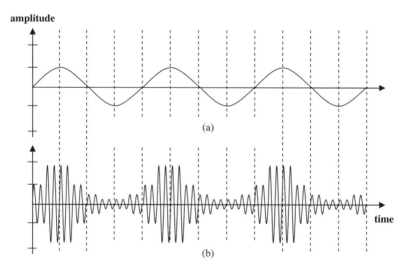

Figure 3.10 Analog-to-analog modulation techniques (a) original signal; (b) amplitude modulation

according to the frequency of the original signal. It is also possible to shift the frequency of a signal to a higher or lower frequency.

Pulse code modulation (PCM), differential PCM and delta modulation are *analog-to-digital* modulation, i.e. digitization, techniques. In PCM, an analog signal is sampled at each sampling interval, S, and the amplitude of the signal for the sampling interval is recorded. The fidelity of the modulation depends on both the length of the sampling intervals and the number of bits used to represent the amplitude of the signal at a sampling interval. The higher the sampling rate, the lower the digitization loss becomes. According to the sampling theorem, when the sampling rate $1/S$ is at least twice as high as the frequency of the most significant signal component, i.e. $2f \le 1/S$, the result of the digitization process contains all the information in the original analog signal.

In differential PCM, instead of the amplitude for each period, the change in the amplitude compared to the previous interval is recorded by using a smaller number of bits. This technique is called delta modulation when only one bit is used to indicate whether the amplitude is lower or higher than the previous sampling interval (Figure 3.11). Since an analog signal can sometimes change more rapidly than it can be represented by delta modulation, digitized data is not always the same as the original signal. The difference between the original and digitized signals is called *digitization noise* or *digitization loss*.

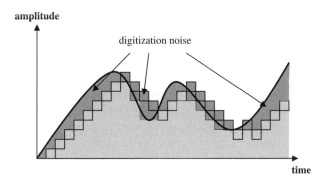

Figure 3.11   Delta modulation

3.1.6 Manchester Encoding

When the clocks at two ends of a link are not perfectly synchronized, the sampling intervals of the sender and the receiver may become different from each other after a long sequence of only 1s or 0s, and the receiver may not be able to distinguish when a bit starts or ends. Therefore, a means of unambiguously determining the start and the end of each bit is needed. Manchester encoding provides this mechanism.

In Manchester encoding each bit period is divided into two equal intervals and both 1 and 0 are transmitted for each bit. This ensures that there is a transition from 1 to 0 or from 0 to 1 in the middle of a bit period. There are two versions of this technique, namely *Manchester* and *differential Manchester* encoding (Figure 3.12). In Manchester encoding, first 0 then 1 is sent to transmit one bit of 0, and first 1 then 0 is sent to transmit one bit of 1. In differential

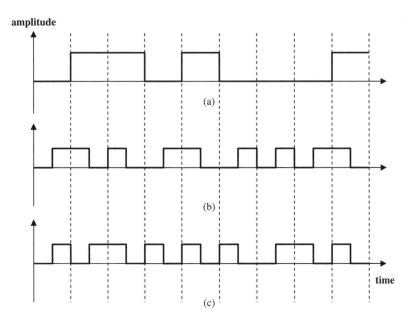

Figure 3.12 Manchester and differential Manchester encoding (a) original signal; (b) Manchester encoding; (c) differential Manchester encoding

Manchester encoding, when the new bit in the data stream is different from the previous bit, there is also a transition in the transmitted stream, otherwise no transition occurs at the end of the bit interval. Note that there is always a transition from 1 to 0 or from 0 to 1 in the middle of every bit interval, therefore ensuring that there is a change in the wave pattern in each bit interval, which helps synchronize the two ends, i.e. the sender and the receiver. However, this is achieved at the expense of 100% overhead. For each bit both 1 and 0 need to be transmitted.

3.1.7 Multiplexing and Duplexing

Another challenge in the physical layer is to share the same link among multiple channels. This is achieved by a process called *multiplexing*, where multiple channels are aggregated into a single link. A multiplexed link is deaggregated into multiple channels at the other end by *demultiplexing*. Note that this is different from multiple access, which we explain in the following chapter, because in multiple access a single channel is shared among multiple terminals. We can divide a given capacity into multiple channels by using multiplexing and demultiplexing techniques, and then use multiple access schemes to share each of these channels among multiple users.

Two well-known techniques for multiplexing are *frequency division multiplexing (FDM)* and *time division multiplexing (TDM)*. In FDM, available spectrum is divided into smaller frequency portions, each of which becomes a frequency channel. In order to avoid interference among the frequency channels, they are also separated by *guard bands*. In TDM, a channel of larger bandwidth is divided into time slots and each slot is assigned to a channel.

To divide the capacity into two directions between two communicating nodes, the same approaches can be used. Some frequency channels can be assigned to one direction while the others can be assigned to the opposite direction by using *frequency division duplexing (FDD)*. Similarly, this can be achieved on a time slot basis by *time division duplexing (TDD)*.

3.2 Advanced Radio Technologies

Advanced radio technologies can increase the capacity for the same bandwidth and reduce the transmission power required for the same transmission range and the same capacity. Most of these technologies are based on better techniques dealing with noise and interference. They reduce the interference by selecting more appropriate frequencies, planning more effective transmission schedules and limiting the transmissions to the intended communications areas as much as possible. They also use frequency, time and space diversity techniques to become more resilient to noise. All these have some implications on security and provide opportunities to develop new and more effective security schemes. New radio technologies also offer more flexible and reconfigurable radios, which may be more susceptible to node tampering, viruses, worms and Trojan attacks.

3.2.1 Directional and Smart Antennas

Directional antennas spatially limit and direct a transmission towards the intended area of communication and reduce the interference in the other directions. This leads to higher capacity. Since they are bigger, more expensive and more sensitive to mobility, they are not considered a suitable technology for mobile ad hoc network nodes but are mostly used for access points, base stations and backbone mesh routers. Two types of directional antenna approaches can be distinguished: *switched beam* and *adaptive* antennas (Figure 3.13)

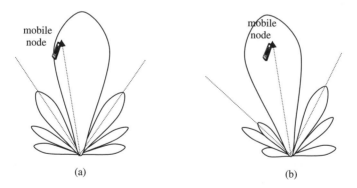

Figure 3.13 Directional antennas (a) switched beam antenna; (b) adaptive antenna

In the switched beam approach, there are fixed beam patterns. The pattern that provides the strongest signal for the mobile node is used. However, the mobile node may not be at the central axis of the main lobe because the beam patterns are fixed. On the other hand, there is no fixed beam pattern in the adaptive antenna approach, where the signal patterns can be dynamically altered towards the mobile node. Therefore, the range and the interference suppression of adaptive antennas are higher than for switched beam antennas.

Smart antennas are advantageous when security is a high priority because emission towards unintended directions is limited and therefore adversaries in those regions cannot receive these transmissions.

3.2.2 Software Radios

Traditional radios are optimized for specific frequencies and protocols, and implemented in hardware. Software radios change this traditional approach in the following ways:

- **Analog-to-digital conversion (ADC) as close to the antenna as possible:** in traditional radios, the received signal is first converted to a lower frequency, called an *intermediate frequency (IF)*. Then the IF signal is filtered for noise, amplified and demodulated. After ADC is complete, digital processing of the incoming signal starts. The closer the ADC is to the antenna, the sooner the digital processing can be applied, which means more components of the radio can be implemented in software. There are important challenges inherent in realizing this. First, fast digital signal processors are needed to accurately sample high frequency analog signals. Second, this also requires linear amplifiers that can amplify wideband signals.
- **Generic hardware:** to run the digital processes, generic hardware instead of specific hardware optimized for a specific frequency and set of protocols is required. This means more cost-effective and flexible but less powerful and time-efficient hardware.
- **Software implementation of the digital processes:** with generic hardware, the radio functions examined in the previous sections, e.g. signal generation, modulation and demodulation, Manchester encoding, frequency hopping, etc., as well as other protocols such as medium access control, encoding, encryption, etc., can be implemented in software. Hence, the radio becomes reconfigurable and the same radio can be used for multiple protocol sets and frequencies. However, they also become vulnerable to viruses, worms and Trojan attacks.

3.2.3 Cognitive Radios

Software radios provide the base for realizing cognitive radios that can observe the available spectrum and choose dynamically the frequency and other parameters at which to operate. A traditional radio runs at a preset frequency channel by using a fixed protocol set in at least the physical layer. A cognitive radio can select a frequency channel and a protocol set based on parameters like availability, capacity and congestion. A high percentage of the available frequencies are usually not in use even when some channels are heavily congested. Cognitive radios can select one of the unoccupied frequencies instead of competing for already congested ones. Hence, a very large capacity can be achieved with no new allocation of spectrum.

Cognitive radios also offer new features for security-related considerations and applications. They can find secure channels and avoid jammed channels. They can also monitor a wide range of channels for various communication technologies, and can detect and listen when there is an ongoing communication on one of them. Therefore, cognitive radios can also be used to set security attacks.

3.2.4 Multiradio/Multichannel Systems

Mesh routers used in wireless mesh networks often use multiple channels and multiple radios. They generally have at least one backbone and one access channel. Apart from these, they often support multiple radio technologies such as IEEE 802.11, Bluetooth and ZigBee to be able to connect various networks. The availability of multiple radios and channels introduces additional challenges, especially in the network and transport layers. Moreover, there is a greater number of points to attack in these systems and therefore defending against security attacks requires more effort. On the other hand, they are more resilient to denial-of-service attacks because there are alternative routes for forwarding the traffic. In addition to this, data can be partitioned and conveyed through multiple channels, which makes eavesdropping more difficult. Multiple routes also help to develop more effective error detection and recovery schemes, as explained in the following chapter.

3.2.5 MIMO Systems

Multiple input, multiple output (MIMO) is a model for multiple antenna communications systems. The MIMO concept is based on multiple antennas and the exploitation of the multipath phenomenon. MIMO algorithms combine both the direct and multipath copies of the received signals, and use multipath copies of a signal to carry more information. Therefore, MIMO increases the spectral efficiency of a wireless communications system by exploiting multipath propagation.

MIMO systems transmit over at least two antennas. In a MIMO system, data throughput and the range increase and bit error rates decrease as the number of antennas increases. The IEEE 802.11n task group uses MIMO as the basis for a WLAN specification that has at least 100 Mbps capacity. 802.11n can be used for wireless mesh networks.

3.3 Review Questions

3.1   There is a communications system that operates in VHF and can provide signal strength 25.5 orders of magnitude higher than the noise level at its maximum range.

(a) What is the total capacity of the system? Assume that all VHF is allocated to this system.
(b) What should be the minimum number of values that each symbol represents to reach this capacity?

3.2   Calculate the propagation latency for a five-hop underwater acoustic network where each hop is 100 meters on average. Assume that there is no processing delay at hops.

3.3   What is the maximum range of a base station that operates at 1.8 GHz channels? Assume that it does not have any transmission power limitations, its antenna height is 30 meters and the receiver is a mobile phone carried by a person. Make your calculations for the worst case scenario.

3.4   What is the path loss at 100 meters in dB for a wave that has a 3 meter wavelength?

3.5   A sensor network operates at 900 MHz. One of the nodes has two routes to send its packets to the collecting node. One of the routes has ten hops of 100 meters. The other route has five hops of 200 meters. Because of the low-lying antenna, the path loss exponent is four. Assume that the nodes do not consume energy for processing the packets to be relayed. Which route is more power efficient?

3.6   Discuss whether or not the systems that operate at higher frequencies are more secure.

3.7   What is the thermal noise for a 30 KHz channel when the temperature is 20 °C?

3.8   What is Doppler fading?

3.9   What is multipath fading?

3.10 Explain the difference between the symbol rate and the data rate.

3.11 Draw the wave pattern that represents the following hexadecimal data stream for QAM-16:

A01E983D

3.12 Compare PCM and delta modulation.

3.13 Write the bit stream generated by Manchester encoding and differential Manchester encoding for the following stream:

010011000111001011

3.14 Discuss the security considerations related to smart antennas.

3.15 Discuss the security considerations related to software radios.

3.16 Discuss the security considerations related to cognitive radios.

3.17 Discuss the security considerations related to multiple channel/multiple radio systems.

4

Medium Access and Error Control

To satisfy the broadband service demand in the mobile and error-prone wireless environment, an efficient medium access control (MAC) scheme is required. MAC schemes have significant impact on the system performance, the system capacity and the hardware complexity. A successful MAC scheme needs to take full advantage of the traffic and network characteristics to fulfill the compelling requirements of WASMs.

MAC is considered a part of the data link layer (DLL) in the Open Systems Interconnection (OSI) reference model (Tannenbaum, 2003). The DLL also covers error and flow control on a link basis. Wireless networks introduce additional challenges for flow and error control. Although these challenges are exacerbated by the mobility and self-configuration requirements of ad hoc networks, as well as the stringent resource and energy constraints of WASMs, error control schemes developed for wireless networks in particular may also be applicable to many WASM applications.

4.1 Medium Access Control

MAC schemes for wireless networks have been extensively studied and numerous MA protocols have been proposed in the literature. In this chapter, we first classify the MA schemes with an holistic approach and examine those commonly used in wireless and tethered networks. Then we explain MAC protocols specifically designed for WASMs.

4.1.1 Generic MAC Protocols

MAC protocols can be broadly categorized into three classes: *contention-based*, *conflict-free* and *hybrid* schemes (Table 4.1). In contention-based schemes, the transmission of a frame, i.e. a chunk of data to be transmitted in the data link layer, is not guaranteed to be successful. The transmission may collide with the transmission of another node. A contention-based MAC protocol resolves collisions once they occur. Conflict-free protocols ensure that the transmission of a frame is not overlapped by the transmission of another frame, i.e. interfered with by another transmission. In conflict-free techniques, after the resources are allocated to a node, they are owned by the node until it does not need them any more and returns them.

Security in Wireless Ad Hoc and Sensor Networks   Erdal Çayırcı and Chunming Rong
© 2009 John Wiley & Sons, Ltd

Table 4.1 Generic MAC schemes

Contention-based	Conflict-free	Hybrid
ALOHA: transmit whenever ready.	FDMA: allocate frequency channels.	PRMA, D-TDMA: contention-based for obtaining channels, and when channels are dedicated, use them conflict-free.
Slotted ALOHA: transmit at the beginning of a time slot.	TDMA: allocate frequency and time slot combinations.	
CSMA: sense the carrier before transmission.	FH-CDMA: frequency hop over several frequency channels.	
CSMA/CD: stop transmission when a collision is detected and try again.		
CSMA/CA: avoid collisions by notifying hidden terminals.	DS-CDMA: spread the data into a larger spectrum.	

During this period, the resources may not be 100% utilized by the node. Hybrid approaches use a contention-based period to obtain the resources for the transmission, followed by a conflict-free period to utilize dynamically allocated resources. Therefore, they offer dynamic allocation and better utilization of scarce resources. It is also possible to classify conflict-free techniques as either *fixed-allocation* or *dynamic-allocation* conflict-free techniques, and to consider the hybrid approach as the dynamic-allocation conflict-free category.

Contention-based techniques are also often called *carrier sense multiple access (CSMA)* based schemes. However, the root of this category is not CSMA but another protocol called *Aloha*. Therefore, one may also call this category the Aloha family. Aloha is a very simple protocol where a node transmits a frame whenever it has one. If the transmitted frame does not reach its destination, it is transmitted again. This is repeated until the successful delivery of the frame. Aloha has very low overhead and reduces delay when there is low traffic load distributed uniformly. However, when the data traffic is heavy and bursty, the collision rate may become too high in Aloha.

Aloha can be enhanced by introducing time slots. Nodes are synchronized and they start transmitting a frame only at the beginning of a slot. This ensures that when a transmission can be started collision free, it can be completed successfully. This method reduces the collision rate at the expense of synchronization of the neighboring nodes and is called *slotted Aloha*.

CSMA is the next stage in the evolution of the Aloha family. In CSMA, the carrier is first sensed to ensure that there is not an ongoing transmission, and the transmission is started if the carrier is free. Although CSMA reduces the collision rate considerably compared to Aloha, there is still the probability that transmitted frames may collide with other frames because two or more nodes can sense the carrier at the same time and mistakenly perceive it as free to transmit. Moreover, since there is latency, a node may sense a carrier to be free even though another node has already started a transmission. Therefore, the collision probability P_c in CSMA is a function of the number n of nodes sharing the carrier, the average latency d and the average frame length l, as shown in Equation (4.1). The higher the number of nodes and average latency, the higher the collision probability becomes. On the other hand, as the average frame length increases, the collision probability decreases.

$$P_c \approx d \times {}^n\!/_l \qquad (4.1)$$

There are three different versions of CSMA: *persistent CSMA, nonpersistent CSMA* and *p-persistent CSMA*. The difference between these three versions is based on the way that they behave when a carrier is detected busy. In persistent CSMA, a node that detects a carrier busy waits until the carrier becomes idle, and when it is idle the node starts transmitting its frame. However, the collision probability may increase in this case because there may be other nodes that also detect the carrier busy and wait until the ongoing transmission is completed. To tackle this, nonpersistent CSMA waits a random period when it detects the carrier busy, and then senses the carrier again before starting transmission. P-persistent CSMA applies to slotted carriers. In this scheme, when a node detects the carrier idle, the probability that it starts transmission is p. The node may not start its transmission with probability $q = 1 - p$, even though it detects the carrier idle, and may wait for the next slot, when it repeats the same algorithm.

Collisions can be detected by transmitting nodes when they are tethered. *CSMA/collision detection (CD)* exploits collision detection to further improve the performance of CSMA. In CSMA/CD the transmitter quits transmission as soon as a collision is detected. In the MAC layer, the Ethernet uses the CSMA/CD algorithm, where a node that detects a collision first waits a random period and then repeats the algorithm. If another collision occurs, the node stops transmission one more time and waits another random period. However, the mean value for the random period is twice as large as the mean value used for the previous random period. This increase in random periods continues by doubling the previous mean value until the successful transmission of the frame. Therefore, this mechanism is called *exponential back-off*. It implicitly provides an approach reactive to the traffic load.

Wireless nodes cannot send and receive simultaneously by using the same antenna. In addition to this, the hidden terminal problem prevents wireless nodes detecting collisions. Moreover, the data collisions caused by the hidden terminal problem lead to retransmissions, and overhearing data transmissions as a result of the exposed terminal problem consumes power. It is obvious that both hidden terminal and exposed terminal problems result in the unnecessary loss of energy. In CSMA/CA, the node that needs to transmit a message sends a small request-to-send (RTS) message to the receiver. The receiver immediately responds with a clear-to-send (CTS) message. After receiving the CTS, the sender transmits the data message. Both the RTS and the CTS messages have a field that indicates the length of the data message. The nodes hidden to the sender or the receiver hear either an RTS or CTS message and avoid accessing the medium for at least a time period equal to the message length given in the RTS and CTS signals. Therefore, collisions are avoided at the expense of additional messaging. Collision avoidance is explained in more detail in Section 4.1.2. Here we would like to highlight the difference between the collision probability in wireless and tethered networks. When collision avoidance is not applied in wireless networks, the probability P_c that a frame collides with another frame is related to the average latency d between the nodes, the number n of nodes and the average length of frames l, as in Equation (4.2).

$$P_c \approx d \cdot n \cdot l \qquad (4.2)$$

According to Equation (4.2), the longer the frames, the higher the collision probability becomes. This is the opposite of the relation in Equation (4.1).

Conflict-free transmission can be achieved by allocating a channel to a node and ensuring that another node does not access the channel. For this purpose, a channel is viewed from a time, frequency, mixed time–frequency or code standpoint. Hence, a multiple access channel can be shared on a frequency basis by using frequency division multiple access (FDMA), on a time basis by using time division multiple access (TDMA) or on a code basis by using code division multiple access (CDMA) (Figure 4.1).

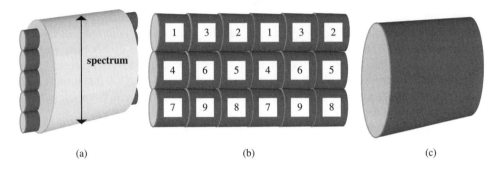

      (a)   (b)   (c)

Figure 4.1 Conflict-free multiple access schemes (a) FDMA; (b) TDMA; (c) CDMA

In FDMA, the spectrum is divided into equal frequency channels, e.g. 30 kHz each, and channels are appropriately spaced. Then, each of these channels is assigned to a single node. TDMA makes use of digital technology. In TDMA, the available spectrum is first divided into larger frequency channels compared to the channels in FDMA. For example, in GSM, each frequency channel is 200 kHz. The frequency channels are further partitioned into time slots, e.g. of 0.5 milliseconds. After this, each node is assigned a specific frequency/time slot combination. CDMA is a spread-spectrum technique where wider frequency channels are used simultaneously by multiple nodes.

There are two CDMA techniques: *frequency hopping (FH)* and *direct sequence (DS)*. An FH receiver and transmitter are assigned N frequency channels for an active call, and they hop over those N frequencies with a hopping pattern known to both of them. For example, a node can hop over 100 channels of 10 kHz each. There are two basic hopping patterns; one is called *fast hopping*, which makes two or more hops for each symbol, the other is called *slow hopping* and this makes two or more symbols for each hop.

DS-CDMA is a multiple access technique where several independent users simultaneously access a channel by modulating signature waveforms, known as *pseudo noise (PN) sequences* or *spreading codes*. This process is called *spreading*. The incoming signal at the receiver is a superposition of such signals. The receiver demodulates and decodes the incoming signal by applying the same spreading code as the sender. This process is called *dispreading*.

The spreading process is illustrated in Figure 4.2, where the data rate is $1/T_d$, and the spreading sequence rate is $1/T_c$. The spreading sequence rate is known as the *chip rate*. Since T_c is 10 times shorter than T_d, the chip rate is 10 times higher than the data rate in this example. When the data are multiplied by the spreading sequence, the resulting signal has the same rate as the spreading sequence. In other words, the original bit stream is

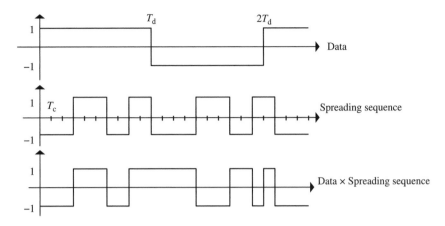

Figure 4.2 Spreading processes

converted to another stream at chip rate. So we have spread our data to a larger spectrum. If we multiply the resulting signal one more time by the spreading sequence, we obtain exactly the original data. This is the best case because there is no other transmission that can interfere with this transmission. When there is interference, we may still dispread the resulting signal and correlate it to the original data unless the interference is too high. Therefore, CDMA has an interference-limited soft capacity. Everything that reduces interference directly increases the capacity. For the interested reader, more detailed information is available at http://www.umtsworld.com/technology/cdmabasics.htm.

Another group of conflict-free MAC schemes is referred to as *token-passing techniques*, where a single (logical or physical) token is passed among the users, permitting only the token holder to transmit; thereby guaranteeing noninterference. Token-ring and token-bus are examples of this class. Since token-based MAC schemes are less and less frequently used in state-of-the-art systems and protocol stacks, we will not elaborate further on them here.

In *hybrid MAC schemes* channels are allocated based on demand, so that a node that happens to be idle does not waste channel resources. These schemes start with a contention-based allocation period when the nodes announce their intention to transmit. Based on these announcements the resources are allocated to the nodes, and those that obtain the resources transmit through the channels allocated to them. Packet reservation multiple access (PRMA) and dynamic TDMA (D-TDMA) are examples of reservation-based protocols.

4.1.2 MAC Protocols for Wireless Ad Hoc, Sensor and Mesh Networks

WASM applications introduce new factors that have an impact on the performance of MAC protocols. As explained in Chapter 2, most of the sensor network protocols are subject to stringent power constraints. Underwater acoustic networks tackle an excessive latency problem. They simply cannot afford RTS–CTS signaling due to high delays. They also need to provide the services over very limited bandwidth. Mesh networks have to satisfy high throughput requirements over a relatively wider bandwidth. Among these constraints, power efficiency has attracted extensive research effort, and power-efficient wireless MAC protocols have been studied intensively since the late 1990s.

A power-efficient wireless MAC protocol should minimize the four sources of energy waste: idle listening, collisions, protocol overhead and overhearing. Table 4.2 provides a chronological list of power-efficient MAC protocols which can be classified into two categories: CSMA-based protocols and TDMA-based protocols. We explain only a subset of these protocols in this book. More detailed information about these protocols can be found in Kumar *et al.* (2006). The protocols in the upper part of the table were originally designed for wireless and ad hoc networks, while those in the lower part are implemented specifically for sensor networks.

Table 4.2 MAC protocols for WASM applications

CSMA-based protocols			TDMA-based protocols			
ALOHA (1970)	MACA (1990)	MACAW (1994)	Distributed TDMA (1996)			
MACA-BI (1997)	PICONET (1997)		Bluetooth (1999)		SMACS (2000)	
PAMAS (1998)	IEEE 802.11 (1997/1999)		NAMA, LAMA, PAMA (2001)		LEACH (2000)	
RBAR (2001)	ARC (2001)	SEEDEX (2001)				
OAR (2002)	S-MAC (2002)	LPL/Pre.Samp. (2002)	ER-MAC (2003)	TRAMA (2003)		EMACS (2003)
T-MAC (2003)	SIFT (2003)	WiseMAC (2003)				
B-MAC (2004)	DMAC (2004)	IEEE 802.15.4 (2003)		LMAC (2004)		
SEESAW (2005)		Z-MAC (2005)		MMAC (2005)		BitMAC (2005)

Multiple access with collision avoidance (MACA) (Karn, 1990) is the first scheme that addresses the hidden terminal and exposed terminal problems. The RTS–CTS signaling scheme comes from MACA and CSMA/CA is based on this protocol. The *MACA wireless (MACAW)* protocol improves the MACA protocol by adding a fourth frame on top of RTS–CTS–DATA frames in order to cope with the unreliability of the wireless channel and to guarantee delivery. When a frame is received correctly, an explicit acknowledgement (ACK) is sent back to the sender node. If the sender node fails to receive an ACK in due time, it retransmits the DATA. The RTS–CTS–DATA–ACK sequence is also the basis for the MAC protocol in IEEE 802.11.

Note that the RTS and CTS frames are assumed to be shorter than the DATA frames, and RTS and CTS signals may collide with the RTS and CTS signals of other stations. However, since RTS and CTS signals are shorter, the cost of their collision is much less than the cost of colliding DATA frames. When DATA frames are shorter than the *dot11RTSThreshold* attribute of the IEEE 802.11 standard, the RTS–CTS signaling scheme has no use.

The RTS–CTS–DATA–ACK sequence is also used in *sensor MAC (S-MAC)* (Ye *et al.*, 2004) and *timeout MAC (T-MAC)* (Dam and Langendoen, 2003), where nodes agree on a common slot structure and switch on and off their radio periodically according to a duty cycle. The S-MAC protocol consists of three major components: periodic listen and sleep, collision and overhearing avoidance, and message passing. Nodes broadcast their sleeping schedule. When nodes receive this schedule from their neighbors, they adjust their own sleeping schedule so that all nodes sleep at the same time. During data transmission periods, nodes exchange data using the RTS–CTS–DATA–ACK signaling scheme. If a node does not have data to transmit or receive, it sleeps. Periodic listen and sleep reduces energy consumption by avoiding idle listening.

T-MAC (Dam and Langendoen, 2003) improves upon S-MAC by introducing an active/sleep duty cycle which is adapted according to the network traffic through a simple timeout mechanism. Nodes communicate by using RTS–CTS–DATA–ACK frames in a similar way to S-MAC. The T-MAC protocol avoids idle listening by transmitting all messages in bursts of variable length and sleeping between bursts. In T-MAC, a node listens and transmits as long as it is in an active duty cycle. An active duty cycle ends when no activation event (data reception, etc.) is detected for a certain time.

In these MAC protocols nodes transmit their frames at the same transmit power level. By applying power control (Cayirci and Nar, 2005; Karlidere and Cayirci, 2006), a node can adjust its transmission power level according to the intended receiver. However, assigning different transmission power levels to different nodes for power control purposes results in the asymmetrical link problem; where node A can reach node B but B cannot reach A, which consequently causes serious collisions. To combat these collisions caused by the asymmetrical link phenomenon, RTS and CTS are transmitted at the highest possible power level, while DATA and ACK are transmitted at the minimum power level necessary to reach the destination in the *BASIC* scheme (Jung and Vaidya, 2002).

Power control MAC (PCM) (Jung and Vaidya, 2002) has been proposed to improve the BASIC scheme. The difference between the PCM and the BASIC scheme is that PCM periodically increases the transmission power level to maximal during DATA frame transmission. In the *PCMAC* scheme (Lin *et al.*, 2003), RTS, CTS, DATA and ACK are transmitted at the minimum power level needed to transmit the frame to the destination, and a separate power control channel is added in order to protect frames from collision at the receiver.

4.2 Error Control

In data transmission in particular, errors cannot be tolerated. They need to be detected and corrected. There are two approaches to tackling transmission errors: *forward error control (FEC)* and *backward error control (BEC)*. Both of the approaches are based on redundant bits being transmitted in each frame. In FEC, the redundant bits suffice both to detect and correct errors. Fewer redundant bits are added to the frames just to detect errors in BEC, where the receiver asks the sender to retransmit the frame when it detects an error. The choice between using FEC and BEC is mainly based on the expected bit error rates. If the medium is so error prone that at least one bit error occurs in every frame of the transmission, BEC is not an applicable technique.

4.2.1 Error Correction

There are numerous FEC schemes such as convolutional and turbo codes. Hamming code is among these schemes. Although Hamming code is not very commonly implemented, other error-correcting schemes are based on the fundamental concepts introduced by Hamming code. Therefore, we explain Hamming code in this section.

The fundamental concept in Hamming code is to differentiate blocks of data from each other such that in the case of a transmission error, the received block can still be correlated to the original data. This can be achieved by adding *redundant* bits called *check bits* to the *data blocks*. The minimum number of bits that a data block has different from the other blocks is called the *Hamming distance*. The Hamming distance indicates the number of bit errors that can be corrected. For example, if the Hamming distance is three, when one bit in a block changes during transmission, the received block resembles the original more than the other possible blocks because it has one bit different from the original block but at least two bits different from the others. Therefore, a single bit error can be corrected.

Let's assume that we partition data into blocks of two bits. There can be four different combinations with two bits. Now we use three check bits for every two-bit data block to correct the transmission errors. In this case we can create a codeword for every two-bit data block which differs from the others in at least three bits, as shown in Table 4.3. When one of these five bits is changed in any symbol, the result is only one bit different from the original but at least two bits different from the others. For example, to transmit 00, five bits, i.e. 00000, are sent. If one of these bits is received erroneously, e.g. 00010, the received codeword is not one of the valid data and check bit combinations shown in Table 4.3. However, its distance to 00000 is only one, while it has at least two bits different from the other valid codewords. Therefore, assuming that there can be an error in only one bit, we can conclude that the transmitted block is 00 after removing the redundant bits in the received block. When more than one bit is received erroneously, we cannot correlate it to any valid codeword. For example, 01001 is not a valid codeword and has a Hamming distance of 2 both from 00000 and 01111. This can happen when two bit errors occur in transmitting either of these codewords. Therefore, we can detect

Table 4.3 Data blocks and codewords

Original data block	Data and check bits
00	00000
01	10101
10	11010
11	01111

but not correct this error. In the case of a higher number of bit errors, this scheme may not detect the error or may even correlate the received data to a wrong codeword.

Hamming code is a block coding system that follows this basic concept. With m check bits, the minimum Hamming distance is three and single-bit errors in data blocks of $2^m - 1$ bits long can be corrected. The check bits are inserted at positions equivalent to the powers of two, starting from the least significant bit. This ensures that when their position in the codeword is expressed in binary form, only the bit at the place of their relative position within the check bits is 1, and the rest is 0.

For example, let's assume that we use four check bits. They are inserted at locations 2^0, 2^1, 2^2 and 2^3 in the transmitted block. These are represented as 0001, 0010, 0100 and 1000 in binary form respectively. These bit strings are called *position strings* and their length is equal to the number of check bits. The maximum number of bits in a data block including the check bits cannot be longer than $2^m - 1$ because that is the highest possible value that can be represented by m bits.

In Hamming code procedure, data bits are first inserted into the positions not for the check bits. Then the position strings of all '1' bits in the data block have an explicit or (XOR) operation applied to them. Each bit in this result is the value for the related check bit. For example, if the result is 1001, this indicates that the most and least significant check bits are 1 and the others are 0. After this, if we XOR the position strings of all the '1' bits, including the check bits, the result should be equal to 0 unless there is a transmission error. If there is a single-bit error, the result of this operation gives the position string for the bit changed during transmission.

In Example 4.1, the bits in positions 5, 9, 11 and 15 are 1. If we XOR the position strings of these bits, the result is equal to 8, which indicates only the most significant check bit is 1

Example 4.1

```
number of check bits m: 4
maximum block length: 2⁴-1 = 15
the data block: 10001010010
```

Pos.	15	14	13	12	11	10	9	8	7	6	5	4	3	2	1
Pos. string	1111	1110	1101	1100	1011	1010	1001	1000	0111	0110	0101	0100	0011	0010	0001
Data	1	0	0	0	1	0	1		0	0	1		0		
Check								1				0		0	0
Block	1	0	0	0	1	0	1	1	0	0	1	0	0	0	0

```
          1111
          1011
          1001
          0101
          1000
```

and the other check bits are 0. Based on this, the *block* shown in Example 4.1 is transmitted. Assume that the bit in position 7 is changed during transmission. When we apply the XOR operation to the bit position string for 7, the result will obviously be equal to seven, which indicates that the seventh bit has been changed during the transmission.

4.2.2 Error Detection

The simplest technique for detecting an error is adding parity bits to the transmitted data. Parity bits introduce very low overhead and do not require much computing. However, they may not detect burst errors. When two or more bits are changed during transmission, they may cancel each other and the receiver may not detect the fact that the frame is garbled. Therefore, another technique called the *cyclic redundancy check (CRC)* is often used for error detection.

In this technique, first of all a bit string called the *generator polynomial* is determined. It is called the generator polynomial because CRC is treated as a polynomial operation where input and generator bit strings are represented as polynomials with coefficients of 0s and 1s. A generator polynomial is used to generate *checksums*, which are appended at the end of frames. The receiver checks the input by using the same generator to detect transmission errors. There are two rules in selecting generator polynomials:

- they must be shorter than the frame length;
- they must start and end with 1.

There are also other desirable features for generator polynomials:

- we can make $x+1$ a factor in the generator polynomial, i.e. the last two bits of the generator polynomial are 11.
- we can select a generator polynomial that does not divide x^m+1, i.e. is not a factor of x^m+1, where m is the maximum frame length.

The algorithm is based on a basic property of the division operation. If the remainder in a division operation is subtracted from the dividend, the result becomes divisible by the divisor. The CRC follows this property in the following three-step algorithm to generate the checksums.

Let's assume that we have a frame of i bits, $F(x)$, and a generator polynomial of k bits, $G(x)$.

1. Append $k-1$ 0s at the end of the frame. The resulting polynomial is $M(x) = x^{k-1} \times F(x)$ and has a length of $m = i+k-1$.
2. Divide $M(x)$ by $G(x)$. Note that this is a polynomial division, i.e. modulo 2 division, where there are no carries for addition and no borrows for subtraction. The remainder is $R(x)$.
3. Subtract $R(x)$ from $M(x)$ by using modulo 2 subtraction. The result is the frame to be transmitted, $T(x)$ of length m.

The destination node receives $T(x)$ and divides it by $G(x)$. If $T(x)$ is divisible by $G(x)$, i.e. there is no remainder, this indicates the transmission is error free. The CRC ensures the following:

- all single-bit errors are detected;
- when $x + 1$ is a factor in the generator polynomial, all errors that invert odd number of bits are detected;
- when $G(x)$ does not divide $x^m + 1$, all isolated two-bit errors are detected;
- all burst errors shorter than k are detected.

In summary, when the generator polynomial is selected carefully, the probability that the CRC cannot detect an error is very low. There are also practical hardware solutions to implement the algorithm. Therefore, it is used in many standards and protocol stacks, such as IEEE 802.

Example 4.2

```
Generator polynomial G(x): 1011, k = 4
Input frame F(x): 1110010100, i = 10
```

> **Step 1:** Append k-1 0s to the input frame
>
> Input frame $M(x)$: 1110010100000, $m = 13$
>
> **Step 2:** Divide $M(x)$ by $G(x)$
>
> ```
> 1110010100000
> 1011
> 0101010100000
>  1011
>  000110100000
>     1011
>     011000000
>     1011
>     01110000
>      1011
>      0101000
>       1011
>       000100
> ```
>
> Remainder $R(x)$: 100
>
> **Step 3:** Subtract $R(x)$ from $M(x)$
>
> Transmitted frame $T(x)$: 1110010100100, $m = 13$

$T(x)$ is divisible by $G(x)$. Therefore, if there is no transmission error, $G(x)$ can divide $T(x)$ at the receiver, which indicates successful transmission.

4.3 Wireless Metropolitan Area Networks

4.3.1 IEEE 802.16

The Wireless MAN standards for broadband wireless access (BWA) in metropolitan area networks (MANs) have evolved since 2001. Initial work denoted IEEE 802.16, IEEE 802.16a
and IEEE 802.16c was consolidated into IEEE 802.16-2004, which is also known as IEEE
802.16d. It defines both the medium access control layer and the physical layer specifications
for fixed broadband wireless access (BWA) systems (ANSI/IEEE, 2004). The amendment
IEEE 802.16e-2005 was later published to cover both fixed and mobile operations (IEEE,
2005b).

The standard defines multiple physical layer specifications for different spectrum bands,
i.e. profiles. Single-carrier modulation is used in the 10–66 GHz bands for fixed, point-
to-point and line-of-sight communications between base stations. For frequencies below
11 GHz, single-carrier or OFDM or OFDMA is employed. Non-line-of-sight communications between a base station and multiple fixed subscriber stations occur in the 2–11 GHz
bands, whereas mobile stations employ frequencies in the 2–6 GHz bands. The physical
layer supports both time division duplex (TDD) and frequency division duplex (FDD)
operation.

As opposed to multiple physical layer specifications, a single MAC layer is provided by the
standard. The MAC layer primarily supports point-to-multipoint (PMP) architecture. Additionally, optional mesh topology is provided below the 11 GHz spectrum. MAC for PMP
topology is a conflict-free scheme where the subscriber station is assigned a dynamic time slot
by the base station. Downlink traffic employs time division multiplex (TDM) while TDMA is
used for the uplink transmissions. In mesh mode, all stations have direct links with their one-
hop neighbors. Therefore, they are required to broadcast their schedules and to coordinate
with their neighbors using a three-way handshake mechanism.

4.3.2 WiMAX

Worldwide interoperability for microwave access (WiMAX) is a seal which certifies equipment as interoperable with other WiMAX-certified equipment. In addition to this, it shows
conformance with the IEEE 802.16 and European Telecommunications Standards Institute (ETSI) high performance radio metropolitan area network (HiperMAN) standards. The
WiMAX Forum (http://www.wimaxforum.org), containing more than 400 member companies, carries out the certification process. The tests for conformance and interoperability are
currently achieved in three labs located in Korea, Spain and China which are designated by
the WiMAX Forum.

Currently, the certification process does not cover all the spectrum bands. Only the key
frequencies that are demanded by the service providers and the equipment manufacturers are dealt with. Products for fixed wireless applications are certified in the 3.5 and
5.8 GHz bands whereas 2.3, 2.5 and 3.5 GHz bands are used for mobile applications. As
the allocation of spectrum evolves, the WiMAX Forum has formed the Spectrum and Reg-
ulatory Database to provide current data to its members regarding worldwide licensing
activities.

As an alternative to cable and DSL in the last-mile BWA, the WiMAX Forum has certified that fixed systems can reach up to 40 Mbps per channel in a radius of 3–10 kilometers.

On the other hand, mobile applications can provide a maximum capacity of 15 Mbps within 3 kilometers.

4.4 Wireless Local Area Networks

4.4.1 IEEE 802.11

The physical (PHY) layer and MAC layer specifications for wireless local area networks are defined in this standard (ANSI/IEEE, 1999). The standard defines two modes of operation: *point coordination function (PCF)* and *distributed coordination function (DCF)*. DCF is aimed at ad hoc networks.

The DCF of the IEEE 802.11 standard is mainly built on MACAW, following the carrier sense multiple access with collision avoidance (CSMA/CA) competition mechanism composed of a four-frame RTS–CTS–DATA–ACK handshake to realize a data transmission between the transmitter and the receiver. IEEE 802.11 performs both physical (at the air interface) and virtual carrier sensing (at the MAC layer). Physical carrier sensing detects activity in the channel via relative signal strength from other sources. Virtual carrier sensing is achieved by sending MAC protocol data unit (MPDU) duration information for each frame in the header of the RTS/CTS and DATA frames. A duration field indicates the amount of time required to complete frame transmission. A local network allocation vector (NAV) is updated with the value of other terminals' transmission durations. Using the NAV, a node's MAC knows when the current transmission ends. The NAV is updated upon hearing an RTS from the sender and/or a CTS from the receiver, so the hidden terminal problem is avoided.

The carrier-sensing range of a node can be divided into two areas: the *transmission range* and the *carrier sensing zone*. The transmissions of the nodes in the transmission range can be received and their content can be read. On the other hand, the transmissions of the nodes in the carrier sensing zone do not have enough SNR to enable their content to be read but the transmissions can be sensed. The nodes in the transmission range set their NAVs according to the received RTS or CTS signals, while the nodes in the carrier sensing zone set their NAVs for the EIFS (extended interframe space) duration, which is quite a long timer, since they cannot decode the frame. A channel is considered to be busy if either the physical or virtual carrier sensing mechanism indicates this to be so, and access is deferred until the current transmission ends.

IEEE 802.11 has various versions, as listed in Table 4.4. The main differences between these versions are in the physical layer specifications. They all follow the CSMA/CA scheme in the MAC layer.

Table 4.4   IEEE 802.11 versions

Protocol	Op. frequency	Maximum data rate	Range
802.11	2.4–2.5 GHz	2 Mbit/s	Not specified
802.11a	5.15–5.35/5.47–5.725/5.725–5.875 GHz	54 Mbit/s	75 meters
802.11b	2.4–2.5 GHz	11 Mbit/s	100 meters
802.11g	2.4–2.5 GHz	54 Mbit/s	75 meters
802.11n	2.4 GHz or 5 GHz bands	54 Mbit/s	125 meters

4.4.2 Wi-Fi

Wi-Fi is a certification mark given to devices that conform to the IEEE 802.11 family of standards, and that are interoperable with other Wi-Fi certified equipment. The Wi-Fi Alliance (http://www.wifialliance.com), with more than 300 members, carries out the certification process.

Wi-Fi provides two modes of operation: infrastructure networks with one or more access points (APs) and ad hoc networks without APs. The traffic flow in an infrastructure network occurs in a star topology between APs and stations. Therefore, an access point periodically transmits its service set identifier (SSID), which is used by a station to determine the AP it will connect to. On the other hand, peer-to-peer communication is employed in the ad hoc mode of operation.

Wi-Fi-certified devices are validated for either 2.4 GHz or 5 GHz ISM bands, or for dual band. They can operate in one of 14 channels depending on local regulations and user preferences. The PHY layer in Wi-Fi employs a direct-sequence spread spectrum (DSSS) transmission system. The data transmission rate can be as high as 11 Mbps or 54 Mbps, and decreases as the signal quality gets poorer. Basic MAC functionality must conform to the CSMA/CA scheme described in the IEEE 802.11 standard.

4.5 Wireless Personal Area Networks

4.5.1 IEEE 802.15.1

The IEEE 802.15.1 standard defines both the PHY and MAC layer specifications for communications in wireless personal area networks (WPANs) (IEEE, 2005a). Devices in a WPAN are assumed to form a piconet. One of these devices is known as the master while the others are called slaves. Slaves are allowed to communicate only with the master node. At most seven slaves can be active at a time while 255 inactive slaves can be waiting for activation by the master. The master node provides the synchronization clock as well as the frequency hopping pattern for the piconet. Frequency hopping spread spectrum (FHSS) is employed in 79 channels at the 2.4 GHz band. The address and the clock of the master node are used together to determine the hopping pattern.

More than one piconet can reside in the same region as long as they use different channels and their master nodes are different. A device can belong to multiple piconets using time division multiplexing (TDM). These piconets are said to form a *scatternet*. Routing in the scatternet is outside the scope of the standard.

The standard categorizes the physical channels into four. An inquiry scan channel is used for device discovery. Connection between discovered devices is established using a page scan channel. Communication between connected devices takes place on either a basic piconet channel or an adapted piconet channel. The access code transmitted at the start of a packet determines the type of physical channel used. A device can operate on all types of channels using TDM as long as only one type of channel is used at a time.

The medium access scheme differs depending on the type of physical channel employed. The basic piconet channel consists of time slots where each slot is associated with an RF frequency in the hopping pattern. Synchronization and time slot numbering are based on the master node's clock. The transmissions of the piconet master can start only in even-numbered slots and may take up to five slots. Slaves' transmissions are controlled by the master.

Each packet from the master node defines how the slaves will respond. Inter-slave communication does not occur on the basic piconet channel. The adapted piconet channel has the same properties as the basic piconet channel except for two issues: first, the responses from the slaves employ the same frequencies as the master's; second, the frequency-hopping pattern may include less than 79 frequencies and mark the others as unused. Communications in both the inquiry scan channel and the page scan channel are based on a point-to-point connection. A device either sends a (page scan or inquiry) request or listens for a response through all (inquiry or page scan channel) frequencies randomly. Devices remain passive in both channels until a request is received and they use a slower hopping rate than the rate of the piconet channels. On the other hand, the requesting devices employ a faster rate and, hence, cover all frequencies in a short time.

4.5.2 Bluetooth

IEEE 802.15.1 is a standard based on Bluetooth. Actually, the initial version of IEEE 802.15.1, which was approved in 2002, adopted physical and MAC layer protocols from Bluetooth v. 1.1. Therefore, they are almost identical in the bottom part of the protocol stack. Bluetooth is an older standardization effort (it was started in 1994) by the industry, with a wider scope with respect to the protocol layers.

In this section we do not explain all the protocols in the Bluetooth stack but define Bluetooth profiles, which are standardized interfaces between Bluetooth devices. Every Bluetooth device must be compatible with the Bluetooth profiles for its application area. Therefore, the following list of Bluetooth profiles is also the list of application areas that Bluetooth is designed for:

- Advanced Audio Distribution Profile (A2DP)
- Audio/Video Remote Control Profile (AVRCP)
- Basic Imaging Profile (BIP)
- Basic Printing Profile (BPP)
- Common ISDN Access Profile (CIP)
- Cordless Telephony Profile (CTP)
- Device ID Profile (DID)
- Dial-up Networking Profile (DUN)
- Fax Profile (FAX)
- File Transfer Profile (FTP)
- General Audio/Video Distribution Profile (GAVDP)
- Generic Access Profile (GAP)
- Generic Object Exchange Profile (GOEP)
- Hands-Free Profile (HFP)
- Hard Copy Cable Replacement Profile (HCRP)
- Headset Profile (HSP)
- Human Interface Device Profile (HID)
- Intercom Profile (ICP)
- Object Push Profile (OPP)
- Personal Area Networking Profile (PAN)
- Phone Book Access Profile (PBAP)

- Serial Port Profile (SPP)
- Service Discovery Application Profile (SDAP)
- SIM Access Profile (SAP, SIM)
- Synchronization Profile (SYNCH)
- Video Distribution Profile (VDP)
- Wireless Application Protocol Bearer (WAPB)

As implied in this list of profiles, Bluetooth is mainly intended for the cordless connection of personal devices to each other, or peripheral devices to them.

4.5.3 IEEE 802.15.4

Another IEEE standard for WPANs is aimed at devices with very limited battery consumption requirements like interactive toys, smart badges, remote controls and home automation. It defines the PHY layer and MAC layer specifications for low data rate wireless connectivity in the personal operating space of 10 m (IEEE-SA Standards Board, 2003).

The PHY layer specifications depend on local regulations and user preferences. A data rate of 250 Kbps at 16 channels is supported in the 2.4 GHz band worldwide. In Europe and North America, 20 Kbps at one channel and 40 Kbps at 10 channels are defined in the 868 MHz and 915 MHz frequency bands respectively. DSSS with BPSK modulation is employed in 868/915 MHz PHY whereas 2.4 GHz PHY employs DSSS with O-QPSK modulation. Energy detection (ED), link quality indication (LQI) and clear channel assessment (CCA) are some of the tasks implemented by the PHY layer.

The standard defines two types of device in WPANs. Full function devices (FFDs) can talk to any other device, and can serve as a PAN coordinator or a device. On the other hand, reduced function devices (RFDs) are allowed to communicate only with FFDs. If there are few FFDs, the devices are organized in a star topology. Otherwise, peer-to-peer topology is employed.

The coordinator can transmit beacon frames that form the 16-slot superframe structure. Both contention-free and contention-based access are possible during a superframe. The contention-free period (CFP) consists of guaranteed time slots (GTSs) dedicated by the PAN coordinator to low-latency applications. The remaining time slots are accessed using the slotted CSMA-CA mechanism. The use of a superframe is optional. Therefore, unslotted CSMA-CA is employed in a no beacon-enabled network as well as in peer-to-peer networks where the peers must either receive constantly or synchronize with each other. Peer-to-peer synchronization is not considered in the scope of the standard.

4.5.4 ZigBee

ZigBee is designed as a reliable, cost-effective, low-power, wirelessly networked monitoring and control product by the ZigBee Alliance, which is an association of companies that works to develop an open global standard for ZigBee and provides certification services for the standard. ZigBee includes IEEE 802.15.4 for the physical and MAC layers. In addition to IEEE 802.15.4, the ZigBee protocol suite also has network and application layer support. ZigBee aims to become the global control/sensor network standard by providing the following features:

- low-cost, low-capacity devices;
- low power consumption;
- a simple and efficient protocol suite;
- scalability for high-density deployment;
- reliable short-range data transfer;
- an appropriate level of security.

In ZigBee there are two categories of physical device: full function devices (FFDs) and reduced function devices (RFDs). An FFD can talk to any other ZigBee device in any topology and can become a coordinator. An RFD is a very simple device that can only communicate with an FFD. Therefore, they can only be slaves in a star topology network. Every ZigBee network requires at least one FFD.

Typical traffic types for ZigBee are periodic for sensor networks, intermittent for control networks and repetitive for low-latency real-time devices. Its network and application layers are designed to support these traffic types in wireless sensor and actuator networks. More details can be found at ZigBee Alliance's website.

4.5.5 WiMedia

WiMedia, like IEEE 802.15.1 and 802.15.4, is intended for WPANs. However, it provides higher transfer rates using ultra-wideband (UWB) technology for multimedia and digital imaging applications, as opposed to 3 Mbps supported by IEEE 802.15.1, which is appropriate for audio streaming.

IEEE 802.15.3 and 802.15.3a were the first standardization efforts for higher rates in WPANs. These two attempts were abandoned in 2006. On the other hand, the previous work on standardization was later taken over by an association known as the WiMedia Alliance (http:www.wimedia.org).

WiMedia is the UWB common radio platform for high-speed and low-power multimedia data transfer in WPANs. The MAC layer and PHY layer specifications for WiMedia are defined by two ISO-based specifications, i.e. ECMA-368 and ECMA-369 (ECMA, 2005). The PHY layer operates in the 3.1 to 10.6 GHz spectrum band using multiband OFDM (MB-OFDM) multiplexing. WiMedia supports data transfer rates of up to 480 Mbps. In addition, it can coexist with other wireless technologies like Bluetooth and Wi-Fi.

The MAC layer includes both reservation-based and contention-based access. Since there is no coordinator device, all devices exchange periodic beacon frames for coordination. Reservation and scheduling information are embedded in the beacon frames. The beacon period (BP) is composed of a variable number of medium access slots and it resides at the start of a superframe. The superframe consists of 256 medium access slots. After the BP, the devices that participated in the reservation send their frames during their slots. The remaining slots are occupied using prioritized contention access (PCA).

PCA is based on four access categories (ACs). A frame is categorized as background, best effort, video or voice, representing the lowest to the highest priorities respectively. For each AC, a different set of parameters is defined, e.g. arbitration inter-frame space (AIFS) and contention window (CW). Using the parameters defined for that category, a CSMA/CA-like access procedure is employed to get a transmission opportunity (TXOP).

4.6 Review Questions

4.1 Discuss the relationship between latency, number of nodes and the collision probability for CSMA. Discuss also the relationship between SNR, collision probability and design factors such as range and number of node limits in local area networks.

4.2 What is the difference between nonpersistent CSMA and p-persistent CSMA? Is the Ethernet a persistent CSMA-based scheme?

4.3 What are the differences between FH-CDMA and DS-CDMA?

4.4 Is S-MAC a power-controlled scheme? How can power control impact on the performance of an ad hoc MAC scheme? What are the additional challenges of power control for ad hoc MAC protocols? Is power control more advantageous for security considerations? Why?

4.5 What is the relationship between WiMedia and IEEE 802.11?

4.6 What are the differences between IEEE 802.11, 802.15.1, 802.15.3, 802.15.4 and 802.16?

4.7 Assume that the following overheads are associated with the error control schemes:

2% for error detection
40% for error correction

What are the maximum bit error rates where these schemes are practical for 500, 1000 and 1500 bit frames? Assume that the receiver sends a 100-bit frame to ask for a retransmission in BEC.

4.8 Answer the following questions for Example 4.1:

(a) Can it be detected if the least significant bit in $T(x)$ is inverted?
(b) Can it be detected if the least significant two bits in $T(x)$ are inverted?
(c) Can it be detected if the least and most significant bits in $T(x)$ are inverted?
(d) Can it be detected if any four bits in $T(x)$ are inverted?

5

Routing

The routing protocol is typically the core of any protocol suite. For example, in the transmission control protocol/Internet protocol (TCP/IP), IP does the routing. Any other protocol in TCP/IP, e.g. ftp, SMTP, TCP, UDP, IGMP, etc, or lower layers, e.g. IEEE 802.11, IEEE 802.3, IEEE 802.16, etc, may or may not be used for transferring data over the Internet. However, every transmission over the Internet has to be encapsulated in an IP packet.

Although our focus is on WASMs, and IP is not always preferred for WASM applications, IP is still important because many WASMs are connected to the Internet. The points where WASMs connect to the Internet can present some exploitation opportunities for security attacks. Also, when a WASM node is communicating with an Internet host, IP packets may be encapsulated in WASM protocols. Moreover, some mesh network applications in particular may be running IP directly. Therefore, we also examine IP in this chapter.

5.1 Internet Protocol and Mobile IP

IP is a link level protocol, i.e. every router demultiplexes the IP header, determines the next router to which the packet will be forwarded according to the parameters in the IP header and encapsulates the IP packet again. The protocols over IP, such as TCP, are end-to-end protocols, which means that their headers are demultiplexed and encapsulated only at the two hosts communicating with each other. IP selects the next router from a routing table which can be updated manually or by the Internet control message protocol (ICMP) in some cases. These routing tables can also be managed by routing protocols such as open shortest path first (OSPF) and border gateway protocol (BGP). Alternatively, IP may apply limited source routing, where the next hop router is already in the header because the source node writes the route to follow into an optional field in the IP header. Source routing can introduce additional security because less secure routers and paths may be avoided by source routing.

5.1.1 IPv4, IPv6 and Security in IP

The functionalities of IP are based on the IP header, which is depicted for IPv4 in Figure 5.1.

Security in Wireless Ad Hoc and Sensor Networks   Erdal Çayırcı and Chunming Rong
© 2009 John Wiley & Sons, Ltd

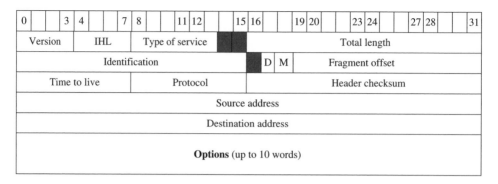

Figure 5.1   IPv4 header

In this book we do not explain each of the fields in this header. The interested reader can refer to Tannenbaum (2003) for the details.

The fixed part of the header is 20 bytes. The header can be extended by optional fields. Therefore, its length is variable and the IP header length (IHL) field gives the length of the header. Since the IHL field is 4 bits, the total length of an IPv4 header can be at most 15 words, i.e. 60 bytes, which means there can be up to 40 bytes of space for the optional fields. There were originally five optional fields, one of which concerned security and was designed to indicate how secret the packet was, and the routers that the source node did not want the packet conveyed through. Although, especially in the military, this may have some use to indicate explicitly which documents are classified, it also helps adversaries to identify which packets have classified content. Consequently, this field is not commonly used and has not proved its usefulness.

IPsec was developed to provide data confidentiality, data integrity and nonrepudiation services for IP. The design is independent from the encryption algorithms and allows multiple granularities, such as protecting all the communications between two routers or only a single TCP connection between two hosts. It has two modes: *transport* and *tunnel* (Figure 5.2).

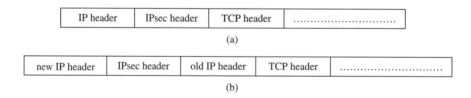

Figure 5.2   IPsec modes (a) transport mode; (b) tunnel mode

The transport mode uses the 'protocol field' in IPv4 to insert an optional header between the IP and TCP headers. The *protocol* field normally indicates the transport layer protocol encapsulated in the IP packet. It may be TCP, user datagram protocol (UDP) or another protocol. When IPsec is used, this field points to an IPsec header.

When tunnel mode is preferred, an IP packet is encapsulated into another IP packet, which applies IPsec, to establish a secure tunnel between two nodes, not essentially the hosts or routers at the end points of a connection (Figure 5.3). Such a tunnel can also be used to aggregate multiple flows into a single stream, which makes traffic analysis more difficult for adversaries.

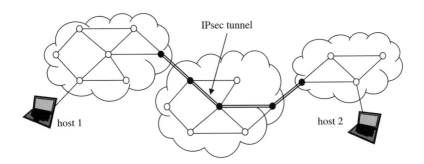

Figure 5.3   IPsec tunnel mode

There are two types of IPsec header: the *authentication header* (AH) and the *encapsulating security payload* (ESP). The AH has the fields depicted in Figure 5.4. The security parameter index (SPI) in this header is the identifier for a *security association* (SA). Although IP is not an end-to-end basis connection-oriented protocol, IPsec requires the establishment of a security association, which is basically an agreement on an SPI that will be valid for a period of time. The sequence number (SN) is a unique number for each IPsec header that has the same SPI. When all the SNs are used for the same SPI, a new SPI is selected, which means terminating the previous SA and establishing a new one. These two fields are for protection against replay attacks. Authentication data is the result of a hash operation over the packet and a shared key, i.e. a hash message authentication code (HMAC). We explain hashing schemes for authentication in detail in Chapter 9. Authentication data ensures the integrity of the IP packet.

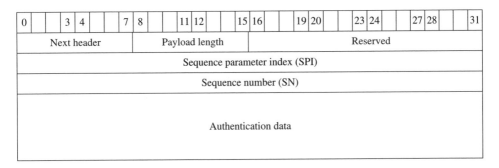

Figure 5.4   Authentication header

The AH is used for data integrity and nonrepudiation services but not for data confidentiality. The ESP header supports data confidentiality and has an initialization vector field for encryption apart from the SPI and SN fields. This initialization vector is used by the selected encryption algorithm.

The IPsec header is inserted between the IP header and the rest of the IP packet by indicating this in the *protocol* field of the IPv4 header. IPv6 has a field called *next header*, which replaces the *protocol* field in IPv4 (Figure 5.5). The header of an IPv6 packet can be extended, and the *next header* field tells which type of header follows the current header. The AH and ESP are therefore header extensions in IPv6.

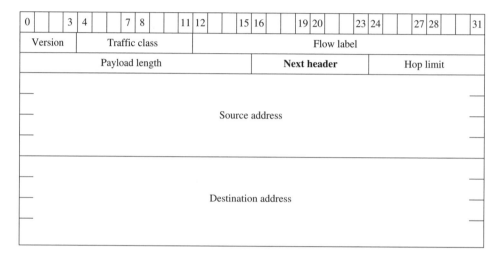

Figure 5.5   IPv6 header

5.1.2 Distance Vector and Link State Algorithms

IP runs in link level and forwards the packets to the next router in the route to the destination. The next router is looked up in a routing table that can be maintained by a routing algorithm. The earlier routing algorithms developed for this purpose were based on a distance vector scheme. The later ones use a link state approach.

In a *distance vector* scheme, each router maintains a table where the next hop router and a metric that indicates the cost of the route through that router are specified to reach every other router in the network. The cost of the route can be determined according to a parameter, such as the number of hops, the probability of congestion, etc. The routers send their tables to their neighbors periodically. When a new routing table is received from a neighbor, the router compares it with its table to see if there is a better route through the neighbor that sends the table. To do so it adds the cost of reaching that neighbor to the costs of the routes in the received table. If the result is less than the value in its own table, it updates its routing table by substituting the next hop id with the id of the neighbor that sends the table and the cost of the route with the result of the calculation for that specific destination.

For example, assume that node *b* in Figure 5.6(a) initially has the routing table in Figure 5.6(b) Then it receives node *e*'s routing table shown in Figure 5.6(c) The cost between

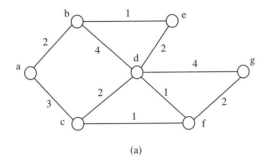

destination	next	cost
a	a	2
c	a	5
d	d	4
e	e	1
f	d	5
g	d	8

(b)

destination	next	cost
a	b	3
b	b	1
c	d	4
d	d	2
f	d	3
g	d	5

(c)

destination	next	cost
a	a	2
c	a	5
d	e	3
e	e	1
f	e	4
g	e	6

(d)

Figure 5.6 Distance vector algorithm (a) example network; (b) initial routing table of node *b*; (c) routing table of node *e*; (d) updated routing table of node *b*

nodes *b* and *e* is one. According to the routing table coming from node *e*, the cost between nodes *e* and *d* is two. Therefore, if node *b* sends its packets destined for node *d* via node *e*, the total cost is three. However, in node *b*'s table the cost between nodes *b* and *d* is four, which is higher. Therefore, node *b* updates its table such that it forwards its packets destined for node *d* via node *e*. Similarly, it also updates the records related to nodes *f* and *g*.

The distance vector algorithm introduces a problem called *count-to-infinity*, this is where an update about a router that goes down is not disseminated effectively. To see this problem, trace the algorithm for the case where every node in Figure 5.6(a) has a perfect routing table, node *d* goes down, node *e* discovers this and nodes *b* and *e* exchange their routing tables in the following order: *b, e, b, e* and *b*. The interested reader should refer to Tannenbaum (2003) to learn more about the count-to-infinity problem.

The *link state* scheme does not have the count-to-infinity problem. In the link state scheme, every node discovers the delay to its neighbors by ECHO messages and disseminates this information to the network by flooding. There are some minor challenges related to this flooding procedure, such as loops and delayed packets that carry neighbor information. However, there are also solutions for them. By collecting these data, every node has a complete picture of the network. Then, one of the shortest path algorithms can be run to determine the best routes for every other router in the network.

The *routing information protocol (RIP)* was the original routing algorithm for ARPANET and the Internet, and it is based on the distance vector scheme. Later algorithms, such as *open shortest path first (OSPF)* and *intermediate system–intermediate system (IS–IS)*, are link state algorithms.

Note that various kinds of attack can be developed based on false update or routing table messages. For example, a malicious router can make itself a very attractive node for all the other routers by disseminating false routing tables or link data. This becomes especially threatening for routing in ad hoc networks. We will examine the types of attack that can be developed against routing schemes in the following sections and in Chapter 8.

5.1.3 Internetworking

The Internet is the network of many autonomous networks. Each autonomous network is made up of several border and gateway routers as well as the links between them. Border routers connect the autonomous networks to the Internet. Although the Internet is not owned by anyone and is created by the voluntary joining of autonomous networks, the autonomous networks are owned by institutions, organizations and individuals, and the traffic through or in the autonomous networks may be regulated by the owners of the autonomous networks. For example, the autonomous network B in Figure 5.7 may want to prevent the packets to autonomous network D from being conveyed through it. Therefore, there are some considerations for internetworking that differ from those for autonomous systems, and two different sets of networking protocols are required: *interior gateway protocols (IGPs)* for routing inside an autonomous network and *exterior gateway protocols (EGPs)* for routing among the autonomous networks.

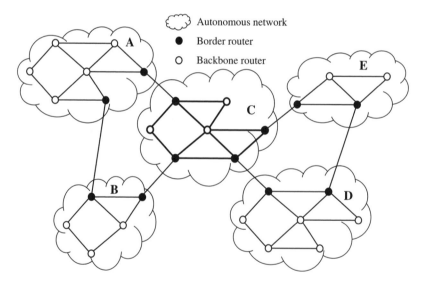

Figure 5.7 An example of an internet – a network of networks

RIP, OSPF and IS–IS are designed mainly as IGPs. The *border gateway protocol (BGP)* is the most well-known EGP. By using the BGP, policies such as, the following may be applied to routing decisions:

- packets to or from autonomous network D cannot go through autonomous network B;
- autonomous network E does not allow any transit traffic.

These policies can be applied for security considerations. Therefore, EGPs may provide some security in the networking layer.

5.1.4 Multicasting, Geocasting, Anycasting and Broadcasting

The simplest and most common type of communication is between two random hosts, but there are also other types of communication pattern where one node transmits a packet to multiple nodes, i.e. multicasting. The most trivial approach for multicasting is to generate a separate copy for each destination and to transmit them separately. This is also the most costly approach. Although the destinations are different, the routes between the source and destinations may follow the same path up to a certain router. Therefore, instead of sending multiple copies over these paths, transmitting a single copy for the packets that follow the same path can be more cost effective. Although this is a challenging issue, there are solutions in the literature (Obraczka, 1998).

The most basic form of multicasting is between a source and a set of random destination nodes. When the destination nodes represent all the nodes in a region that can be geographically defined, this is called *geocasting*. For example, the address for geocasting may be one of the following:

- the nodes on Floor A or in Room B;
- the nodes in City Q or Country R;
- the nodes in the region between Coordinates X and Y.

Sometimes the same packet can be sent to many destinations, but if it is received by only one of them, that suffices. This is called *anycasting*.

Finally, a *broadcast* packet is sent to every other node in a network. Multicasting and broadcasting introduce additional security weaknesses because the probability that a multicast or broadcast packet is eavesdropped is higher when compared to sending a packet to a single destination.

5.1.5 Mobile IP

IP was not designed for mobile networks. It is based on the routing information available in routers. In IP, when a host changes its location, and this location change also implies changes in some routes towards it, the routing tables of routers on these changed routes have to be updated. A mobile device can travel between continents. Therefore, some mobility patterns may require updating all the routers in the Internet. As a result, mobility is too costly for IP, and Mobile IP (Figure 5.8) has been designed mainly to tackle this challenge.

Two important components of mobile IP are *home* and *foreign agents*. Every mobile host has a home agent and an IP address assigned by the home agent. Apart from home agents, foreign agents assign temporary IP addresses called *care-of-addresses* to the visiting mobile hosts. To do this, foreign agents *advertise* their services. When a mobile host is away from its home network, it first waits for these advertisements and sends a registration request to the foreign agent when it receives one. The foreign agent that receives this request assigns a care-of-address to the visiting host and informs the visitor's home agent of the assigned care-of-address. Alternatively, a visiting node that cannot receive any advertisement may transmit

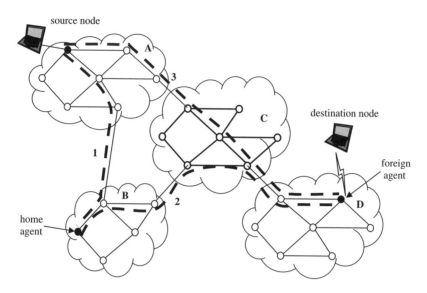

Figure 5.8   Mobile IP

a hello packet to find a foreign agent. Assignment of a care-of-address is called *binding* and it should be refreshed periodically. Otherwise it will be cancelled. Hence, a visitor that leaves without deregistration does not block an address unnecessarily.

If a host sends a packet to a roaming host, this packet is delivered to its home agent, as shown by Step 1 in Figure 5.8. The home agent forwards this packet to the care-of-address of the mobile host, shown in Step 2. After this, the connection is maintained between two nodes through the direct route to the care-of-address.

This scheme introduces additional security challenges. First, an adversary may try to gain access to a foreign agent by using false identification, thus receiving the packets sent to another node. This is solved by authenticating the visiting node before registering it. Second, a malicious node may transmit advertisements and make the mobile hosts in the vicinity register with itself. Another form of attack is again by a malicious node that acts as a foreign agent, but this time it sends false binding information to home agents to make them forward the others packets to a malicious node, i.e. either to itself or another malicious node. As always, a new capability creates new vulnerabilities to be exploited by adversaries.

5.2 Routing in Wireless Ad Hoc Networks

Mobile IP cannot fulfill the requirements for routing in wireless ad hoc networks in which not only the hosts but also the backbone is mobile and multihop wireless connections composed of many links with varying quality of service (QOS) are allowed. Therefore, more adaptive network layer protocols are required. Proactive or reactive approaches can be followed when designing a routing algorithm for ad hoc networks.

A *proactive approach*, often also called a *table-driven approach*, is used by Internet routing algorithms like RIP, OSPF, IS–IS and BGP. In these algorithms, the routers maintain consistent, up-to-date routing information to every other node in the network. Routing tables are updated every time the topology changes. The following are examples of proactive ad hoc routing protocols (Haas and Liang, 1999; Royer and Toh, 1999):

- destination-sequenced distance vector routing;
- cluster head gateway switch routing;
- wireless routing.

In *reactive* techniques, also called *on-demand* techniques, topology maintenance, i.e. maintaining up-to-date topology information in every router, is not continuous but is an on-demand effort. When a new packet needs to be delivered and there is not a valid route to carry out this delivery, a new route is discovered. Examples of reactive techniques are:

- flooding;
- ad hoc on-demand distance vector routing (AODV);
- dynamic source routing (DSR);
- temporarily ordered routing;
- associativity-based routing;
- signal stability routing.

A route may be unnecessarily updated many times before it is used in a proactive approach. On the other hand, the cost of route discovery every time a route is needed may be higher than the cost of maintaining an always up-to-date, consistent view of the network. This depends on the traffic generation and topology change rates. For contemporary wireless ad hoc network applications, reactive techniques such as AODV and DSR are preferred.

5.2.1 Flooding and Gossiping

In *flooding*, each node receiving a packet repeats it by broadcasting unless a maximum number of hops for the packet is reached or the destination of the packet is the node itself. Flooding is a reactive technique and it does not require costly topology maintenance or complex route discovery algorithms. However, it has several deficiencies such as:

- *Implosion* – a situation where duplicated messages are sent to the same node. For example, if node A has *n* neighbors that are also the neighbors of node B, node B receives *n* copies of the same packet sent by node A.
- The flooding protocol does not take into account the available resources at the nodes or links, i.e. *resource blindness*.

A derivation of flooding is *gossiping*, where nodes do not broadcast but send the incoming packets to a randomly selected neighbor. Once the neighbor node receives the data, it selects another node randomly. Although this approach avoids the implosion problem by just having one copy of a packet at any node, it takes a long time to propagate the message to all nodes.

5.2.2 Ad Hoc On-demand Distance Vector Routing (AODV)

AODV is an on-demand ad hoc routing scheme that adapts the distance vector algorithm to run on a network with a mobile backbone. In AODV, every node maintains a routing table where there can be at most one entry for a destination. Each entry has fields like the neighbor node to relay an incoming packet destined to a specific node and the cost of the selected route. This is similar to the distance vector algorithm. AODV differs from the distance vector algorithm by its routing table maintenance mechanism. When a node receives a packet, it first checks its routing table to determine the next hop router for the destination in the packet. If there is an entry for the destination, the packet is forwarded to the next hop router. Otherwise a new route is discovered by broadcasting a route request (RREQ) packet. An RREQ packet includes the following fields: source address, request id, destination address, source sequence number, destination sequence number and hop count. The source address is the address of the initiator of the route request.

If a node receives a route request that has the same source address and request id fields as those in one of the previous route request packets, it discards the packet. Otherwise it checks if there is an entry in its routing table for the destination address. If there is, the destination sequence number in the table is compared to the destination sequence number in the route request. If a router has a route for a destination in its routing table, and if it cannot reach the destination through that route, it increments the destination sequence number and sends a route request. Therefore, the destination sequence number indicates the freshness of a route. If a router has an entry for the destination in its table, and the sequence number for the request is smaller than the sequence number for the destination in its table, this means the route known by the router is fresher than the one known by the router that sends the request. In this case the receiver sends a route reply (RREP). The RREP is forwarded back to the source node through the route where the request is received.

Again, this routing scheme introduces new security challenges. A malicious node may send RREP messages for every RREQ and make the other nodes forward their packets towards it. It may then sink the incoming packets, forward them to another adversary or gain unauthorized access to their contents.

5.2.3 Dynamic Source Routing

Another self-forming and self-healing routing protocol for ad hoc networks is dynamic source routing (DSR). It is similar to AODV in that the DSR protocol is also based on 'route discovery' and 'route maintenance' mechanisms and it is a reactive technique. On the other hand, DSR applies source routing instead of relying on the routing tables maintained by the routers.

In DSR when a node has a packet and it does not know the route for the destination, it sends out a 'route request' packet. While this packet is being transferred through the network, all the nodes traversed are recorded in the packet header. A node that knows the route to the destination does not forward the packet further, but appends the route to the route information already accumulated in the packet and returns a 'route reply' packet to the source node. After this, the source node maintains the discovered route in its 'route cache' and delivers the packets to the destination node through the discovered route by using source routing, i.e. the address of each router to visit until reaching the destination is written in the packet header by the source node. If the routing through a previously discovered route fails, a 'route error' message generated by the node that discovers the route failure is sent back to the source node, the failed route is removed from the 'route caches' and a new route discovery procedure is initiated for the destination.

DSR also introduces security challenges similar to those in AODV. On the other hand, the source node controls the nodes to be traversed and this can be advantageous for security because unreliable nodes can be avoided by the source node.

5.3 Routing in Wireless Sensor and Actuator Networks

There are major differences between the conventional ad hoc networks and wireless sensor networks. These are explained in detail in Chapter 2. In particular, the stringent power constraints and scalability requirements of sensor networks, which differ from conventional ad hoc networks, cannot be satisfied by the routing algorithms designed for ad hoc networks, such as AODV and DSR. Therefore, there are many routing algorithms specifically designed for wireless sensor and actuator networks.

We can categorize routing algorithms for sensor networks as *data-centric, cluster-based* or *location-based*. Data-centric algorithms are based on the specifics of data. Schemes like directed diffusion, sensor protocols for information via negotiation (SPIN) and power-aware many-to-many routing fall into this category. Low-energy adaptive clustering hierarchy (LEACH) is an example of a cluster-based sensor network routing algorithm. The minimum energy communication network (MECN) and geographic adaptive fidelity (GAF) are location-based routing algorithms.

5.3.1 Directed Diffusion

In directed diffusion (Intanagonwiwat *et al.*, 2000) the data-collecting node, called the sink, sends out *interest*, which is a task description, to all sensors, as shown in Figure 5.9(a). The

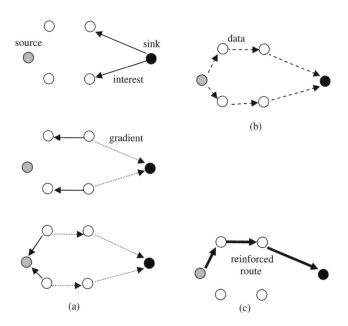

Figure 5.9 Directed diffusion (a) interest dissemination; (b) data dissemination; (c) route reinforcement

task descriptors are named by assigning attribute-value pairs that describe the task. Each sensor node stores the interest entry in its cache. The interest entry contains a *timestamp* field and several *gradient* fields. As the interest is propagated throughout the sensor network, the gradients from the source back to the sink are set up. When the source has data for the interest, the source sends the data along the interest's gradient path, as shown in Figure 5.9(b). The interest and data propagation and aggregation are determined locally. Also, the sink must refresh the interest and reinforce one of the paths when it starts receiving data from the source. When one of the paths is reinforced by the sink, the source node starts sending the packets only through the reinforced path but not all the available gradients. Note that directed diffusion is based on data-centric routing where the sink broadcasts the interest.

5.3.2 Sensor Protocols for Information via Negotiation (SPIN)

SPIN is a family of adaptive protocols based on the idea that sensor nodes operate more efficiently and conserve energy by sending data that describe the sensor data instead of sending the whole data, unless the whole data are explicitly requested (Heinzelman *et al.*, 1999). SPIN has three types of message: ADV, REQ and DATA. Before sending a DATA message, a sensor broadcasts an ADV message containing a descriptor, i.e. meta-data, of the DATA. If a neighbor is interested in the data, it sends a REQ message for the DATA and DATA is sent to this neighbor sensor node. The neighbor sensor node then repeats this process, as illustrated in Figure 5.10. As a result, the nodes that are interested in the data will get a copy.

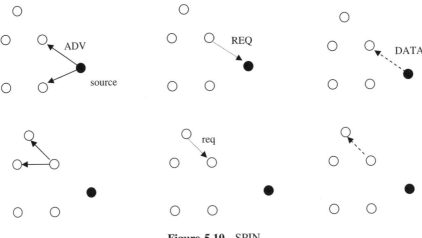

Figure 5.10   SPIN

5.3.3 Low-Energy Adaptive Clustering Hierarchy (LEACH)

LEACH is a clustering-based protocol that minimizes energy dissipation in sensor networks (Heinzelman *et al.*, 2000). The purpose of LEACH is to select sensor nodes randomly as cluster heads, so the high energy dissipation in communicating with the base station is spread to all sensor nodes in the sensor network. The operation of LEACH is separated into two phases: the set-up phase and the steady phase. The duration of the steady phase is longer than the duration of the set-up phase in order to minimize the overhead.

During the set-up phase, a sensor node n chooses a random number between 0 and 1. If this random number is less than a predetermined threshold, t, the sensor node becomes a cluster head. The threshold t is calculated as:

$$t = \begin{cases} \dfrac{P}{1 - P \times \left[r \bmod 1/P \right]} & \text{if } n \in G \\ 0 & \text{otherwise} \end{cases} \qquad (5.1)$$

In this equation P is the desired percentage to become a cluster head, r is the current round and G is the set of nodes that have not been selected as a cluster head in the last $1/P$ rounds. After a node is self-selected as a cluster head, it advertises this to all its neighbors. The sensor nodes receive advertisements and they determine the cluster that they want to belong to, based on the signal strength of the advertisements from the cluster heads. The sensor nodes inform their cluster head that they will be a member of the cluster, and then the cluster head assigns a time slot for every sensor node in which they can send data to the cluster head.

During the steady phase, the sensor nodes can begin sensing and transmitting data to the cluster heads. The cluster heads also aggregate data from the nodes in their cluster before sending them to the base station. After a certain period of time spent in the steady phase, the network goes into the set-up phase again.

5.3.4 Power-Aware Many-to-Many Routing (PAMR)

PAMR (Cayirci *et al.*, 2005) is designed for sensor and actuator networks where sensors send their data directly to the interested actuators, and both power and delay are considered in the routing decisions. In PAMR, actuators register their interest in data with the nodes in the sensor network by broadcasting a *registration message*. A *registration message* includes fields such as *node identification (node_id), actuator identification (actuator_id), echelon, minimum power available (minPA), total power available (totalPA) and task(s)*. Node_id is the identification of the sending node. When an actuator broadcasts a registration message, it initializes the node_id field with its own id, and the nodes that repeat the message update this field. Every node that repeats a registration message replaces the node_id field with its own id. Echelon means the minimum number of hops required to reach a node from an actuator. The totalPA is found by summing up the power available in every *node* along the route. The minPA is the power available in the *node* that has minimum power along the route. A node that relays a registration message adds its power available (ownPA) to the totalPA. It also replaces the minPA field with its ownPA if the ownPA is lower than the minPA value. Before transmitting the registration message, the actuator initializes the echelon and totalPA as 0 and the minPA as the maximum possible PA value.

Sensor nodes do not repeat all the received registration messages. They first check if the registration message is for a new route. A route that meets one of the following criteria is a new route:

- the registration table does not have any entry for the actuator in the registration message;
- the registration table has at least one entry for the actuator, but none of these entries for the actuator is from the uplink node in the registration message;

- the registration table has an entry for the actuator and uplink node in the registration message; however, at least one of the tasks in the message is not indicated in the related registration table entry.

If the registration message is not for a new route, it is checked to determine whether it is about a better route based on the echelon, minPA and totalPA fields. To do this, a parameterized selection function formulated below is used.

Let's assume that we have a sensor node s, and $N = \{n_1, n_2, \ldots \ldots n_n\}$ is the set of uplink nodes in the routing table of s. The general formula for the selection function is:

$$f_i = (w_1 \times \alpha_i) + (w_2 \times \beta_i) + (w_3 \times \phi_i) \qquad (5.2)$$

where w_1, w_2, w_3 are the weighting parameters which satisfy $w_1 + w_2 + w_3 = 1$ and $0 \leq w_1, w_2, w_3 \leq 1$, and

$$\alpha_i = \frac{\sum\limits_{k=1}^{n} e_k - e_i}{\sum\limits_{k=1}^{n} e_k} \qquad \beta_i = \frac{m_i}{\sum\limits_{k=1}^{n} m_k} \qquad \phi_i = \frac{t_i}{\sum\limits_{k=1}^{n} t_k} \qquad (5.3)$$

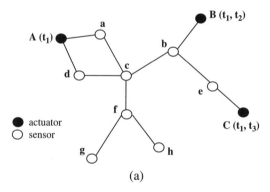

(a)

Actuator Id	Uplink Id	Echelon	Min PA	Total PA	Task (s)
A	a	2	5	5	t_1
A	d	2	4	4	t_1
B	b	2	7	7	t_1, t_2
C	b	3	3	10	t_1, t_3

(b)

Task(s)	Uplink node Id
t_1	a
t_1	b
t_2	b
t_3	b

(c)

Figure 5.11 Example sensor and actuator network (a) example topology; (b) the registration table of node c; (c) the routing table of node c

where e_i is the echelon of the uplink node i, m_i is the minimum power available along the route via the uplink node i and t_i is the total power available along the route via the uplink node i. After calculating f values for all neighboring nodes, the node that has the maximum f value is selected as the uplink node to route an incoming data packet to a specific actuator.

While a registration message is being disseminated in the network, the nodes maintain two tables: a *registration table* and a *routing table*. Examples of registration and routing tables are given in Figure 5.11. The *registration table* of a node is the list of actuators that have registered at least one task to the network. The *routing table* is the list of best uplink nodes for every task and actuator pair available in the registration table. When a new sensed data packet is received, this table is checked for the uplink (i.e. next hop) nodes and the packet is forwarded to every uplink node for the task, i.e. sensor type. The same type of sensed data may be relayed to multiple nodes. Therefore, there may be multiple records for the same task in a routing table. However, there is a single record for every unique uplink node and sensing task pair.

5.4 Review Questions

5.1 Assume that every node in Figure 5.6(a) has a perfect routing table. Then node d goes down and node e discovers this. Trace what the routing tables of nodes b and e become if the distance vector algorithm is used and nodes report their routing tables in the following order: b, e, a, b, e, a.

5.2 How does a link state algorithm differ from a distance vector algorithm? Give two examples of link state algorithms used in the Internet.

5.3 Why do we need to categorize routing algorithms for the Internet as interior and exterior gateway protocols? Why do we need exterior gateway protocols? Elaborate on interior and exterior gateway protocols based on security considerations.

5.4 Why does IP not suffice when nodes are mobile?

5.5 What happens if a mobile node leaves a foreign LAN without deregistering in mobile IP?

5.6 What are the security weaknesses introduced by mobile IP? How do you tackle them?

5.7 Why does mobile IP not suffice for wireless ad hoc networks?

5.8 What are the differences between AODV and DSR protocols? Which one has fewer security weaknesses?

5.9 Compare directed diffusion, LEACH and PAMR for the case where the nodes are mobile.

6

Reliability, Flow and Congestion Control

In OSI, reliability and flow control are tasks for both the data link and transport layers. Error correction and/or detection schemes, explained in Chapter 4, are used to ensure reliability during transmission on a link basis by the data link layer. The transport layer provides reliability as well as flow and congestion control on an end-to-end basis. We elaborate on the challenges and solutions related to transport layer issues for wireless ad hoc networks in this chapter.

6.1 Reliability

The reliability schemes in the transport layer are generally based on retransmission. A data segment is resent by the source node, or sometimes by an intermediate node that has the segment, when it was not delivered successfully on the previous attempt. There are two challenges to achieving this:

- detecting that a data segment has not been delivered successfully, i.e. has been lost or garbled during transmission;
- initiating the retransmission of the segment.

Detection of an unsuccessful delivery can be made either by the source or the destination. If a node that receives the data always acknowledges a successful delivery, this is called *positive acknowledgement*. When the positive acknowledgement scheme is used, if the source node does not receive an acknowledgement for a transmitted segment, it indicates that either the segment has not been delivered successfully or the acknowledgement has been lost. In this scheme the responsibility for detection is with the source node. Alternatively, the destination can notify a failed delivery by a *negative acknowledgement*.

If the data segments are sent sequentially and each segment is assigned a sequence number, the destination can acknowledge multiple segments up to the last received segment by acknowledging the sequence number of the last segment. Transmission control protocol (TCP)

Security in Wireless Ad Hoc and Sensor Networks Erdal Çayırcı and Chunming Rong
© 2009 John Wiley & Sons, Ltd

in the Internet is based on the positive acknowledgement of multiple segments for end-to-end reliability.

The TCP approach for end-to-end reliability can also be used in wireless ad hoc and mesh networks. However, in sensor networks the acknowledgement of segments and retransmission of every lost segment by the source node may be considered too costly due to the stringent power constraints and high number of hops. The traffic patterns in sensor networks are one-to-all, all-to-one and many-to-many. This is also different from the Internet and conventional ad hoc networks. Moreover, the ultimate goal of a sensor network is the detection of events of interest. Since the detection ranges of snodes are often overlapping, the same event is usually reported by multiple sensor nodes. As the successful notification of an event is important in sensor networks, the loss of data packets can be tolerated unless it hinders notification of the event. To overcome these differences, another set of end-to-end reliability schemes is needed for wireless sensor networks.

The *reliable multisegment transport* (RMST) (Stann and Wagner, 2003) scheme is designed to provide end-to-end reliable data packet transfer for directed diffusion. RMST is a selective negative acknowledgement (NACK)-based protocol that has two modes: the *caching mode* and the *noncaching mode*. In the caching mode, a number of nodes along a reinforced path, i.e. the path that the directed diffusion protocol uses to convey the data to the sink node, are assigned as RMST nodes. Each RMST node caches the fragments of a flow. Watchdog timers are maintained for each flow. When a fragment is not received before the timer expires, a negative acknowledgement is sent backward in the reinforced path. The first RMST node that has the lost fragment along the path retransmits the fragment. The sink node acts as the last RMST node, and it becomes the only RMST node in the noncaching mode.

The *pump slowly, fetch quickly* (PSFQ) scheme (Wan *et al.*, 2003) is similar to RMST (Stann and Wagner, 2003). PSFQ comprises three functions: message relaying (pump operation), relay-initiated error recovery (fetch operation) and selective status reporting (report operation). Every intermediate node maintains a data cache in PSFQ. A node that receives a packet checks its content against its local cache and discards any duplicates. If the received packet is new, the TTL field in the packet is decremented. If the TTL field is higher than 0 after being decremented, and there is no gap in the packet sequence numbers, the packet is scheduled to be forwarded. The packets are delayed by a random period between T_{min} and T_{max}, and then relayed. A node goes to fetch mode once a sequence number gap is detected. The node in fetch mode requests the retransmission of lost packets from neighboring nodes.

PSFQ and RMST schemes are designed to enhance end-to-end data packet transfer reliability. The *event-to-sink reliable transport* (ESRT) (Sankarasubramaniam *et al.*, 2003) protocol is a transport layer protocol that focuses on end-to-end reliable event transfer in wireless sensor networks. In ESRT, reliable event transfer is not guaranteed but increased by controlling the event-reporting frequencies of sensor nodes.

In the reliable event transfer approach, an event is defined as the critical data generated by sensor nodes (Tezcan *et al.*, 2004). End-to-end reliable event transfer (EERET) schemes are designed for the delivery of these critical data packets. In most cases, the same critical data are generated by more than one sensor node because sensor nodes are usually densely deployed in sensor networks, and therefore an event may be detected by multiple nodes. An event is successfully transferred to the cnode when at least one packet reporting the event is received by the cnode. For example, nodes *a*, *b*, *c* and *d* in Figure 6.1 detect the same event. Since all four nodes can generate data packets reporting this event, the end-to-end transfer of the event can succeed even if only one of these data packets is received by the cnode.

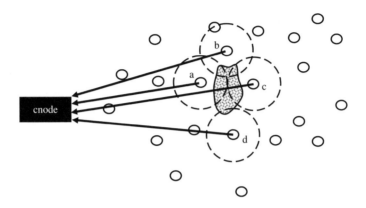

Figure 6.1   End-to-end reliable event transfer

EERET schemes can be categorized into two groups: *non-acknowledgement (NoACK) based* and *acknowledgement (ACK) based* schemes. Three methods, namely event reporting frequency, node density and implicit acknowledgement, can be given as examples of NoACK-based schemes. For the second group, three other schemes, i.e. selective acknowledgement, enforced acknowledgement and blanket acknowledgement, can be listed.

The next two sections explain NoACK-based and ACK-based schemes. NoACK-based schemes are collections of alternative methods of increasing reliability without waiting for end-to-end acknowledgements. In contrast, ACK-based schemes make use of acknowledgement but not as in connection-oriented end-to-end protocols. Acknowledgement is used in different ways to provide reliability.

6.1.1 Non-Acknowledgement-Based Schemes

In this section we cover various schemes in which sensor nodes do not wait for an end-to-end acknowledgement.

6.1.1.1 Implicit Acknowledgement

One method for reliable event transfer in sensor networks is implicit acknowledgement. Implicit acknowledgement makes use of the broadcast characteristic of the wireless channel. When a data packet is sent by a sensor node, it is repeated by its gradient. This implies the successful reception of the packet by the next node. Since it does not need a separate acknowledgement packet, its additional overhead is only due to listening to the media for a time interval. One may argue that this is not an end-to-end scheme but a hop-by-hop technique. This is correct. However, this technique increases the end-to-end reliable event transfer rate.

6.1.1.2 Event-Reporting Frequency

This scheme is used by the ESRT (event-to-sink reliable transport) protocol (Sankarasubrama-niam *et al.*, 2003), which is based on the event-to-cnode reliability model. The level of event delivery reliability is controlled by increasing or decreasing the event-reporting frequency. As the reporting frequency increases, the number of packets sent by a sensor node increases. This decreases the probability that the reported event is lost. However, it also incurs additional power consumption.

One other point is that reporting frequency can be increased until a certain point, beyond which the reliability drops. This is because the network is unable to handle the increased injection of data packets and data packets are dropped due to congestion. Details of this scheme can be found in Sankarasubramaniam *et al.* (2003).

6.1.1.3 Node Density

In sensor networks there are usually multiple nodes that have overlapping sensing regions. Hence, it is possible for multiple nodes to collaborate in detecting the same event. The number of nodes that report the same event has an impact on the end-to-end event transfer reliability. As the number of sensor nodes in critical regions or the number of nodes involved in reporting an event is managed using a network management protocol, the end-to-end reliable event transfer rate can also be controlled. A higher end-to-end reliable event transfer rate can be achieved by increasing the number of nodes involved in a sensing task. A task set concept (Cayirci *et al.*, 2006a) can be used to manage the number of nodes involved in a sensing task.

6.1.2 Acknowledgement-Based Schemes

Although the acknowledgement mechanism is a traditional way of achieving end-to-end reliability, it may not be viable for many sensor network applications for the following reasons:

- most WSN applications have very stringent energy constraints, therefore the overhead of the acknowledgement packets may not be justifiable;
- since some of the packets may not be as critical as others, generating acknowledgements for all packets received may incur unnecessary cost;
- since many sensor nodes may report the same event, acknowledging all of them with a single acknowledgement may be more effective.

More appropriate acknowledgement schemes for sensor networks are introduced in Tezcan *et al.* (2004).

6.1.2.1 Selective Acknowledgement

Since WSNs consist of thousands of sensor nodes, which are densely deployed, waiting for an acknowledgement for each data packet is inappropriate. Instead, each sensor node may activate the acknowledgement mechanism when it detects critical data. There are various ways to determine whether a data packet carries critical data or not. One approach is to use a threshold value. Sensor nodes and the cnode come to an agreement before deployment on a threshold value that depends on the application. Then, a sensor node decides whether the measurement is critical or not by comparing with the agreed threshold value. For example, a temperature sensor may report the temperature periodically. Unless the reported temperature changes by more than a threshold value, the reported data can be accepted as not critical. When a change in excess of the threshold value is observed, this can be accepted as critical. In other cases, lost data can be obtained by interpolation.

Sensor nodes wait for acknowledgements only when critical data are reported. The cnode compares each data packet that it receives with the threshold value and categorizes them as critical or not. When a critical data packet is received, the cnode sends an acknowledgement packet to the sensor node immediately. If the sensor node does not receive an acknowledgement for the packet that contains critical data during a predetermined timeout period, it retransmits the packet.

6.1.2.2 Enforced Acknowledgement

In enforced acknowledgement the basic idea is almost the same as for selective acknowledgement. The difference is that the cnode does not compute whether the received data packet carries critical data or not. Instead, the sensor node computes this before sending the packet and marks the packet if it carries critical data. The cnode sends back an acknowledgement when it receives a packet marked as carrying critical data.

6.1.2.3 Blanket Acknowledgement

Multiple sensor nodes reporting the same event may be acknowledged by a single acknowledgement packet. A single acknowledgement packet can be broadcast to all sensor nodes reporting the same event. For example, the blanket acknowledgement scheme can be used in sensor networks for disaster relief operations where sensor nodes are responsible for reporting humans trapped under rubble. When the cnode acknowledges the presence of a human under rubble, the sensor nodes do not need to worry about whether their particular report has been acknowledged or not (Cayirci and Coplu, in press).

Blanket acknowledgement can also be used in conjunction with selective and enforced acknowledgements to broadcast the acknowledgement packets.

6.2 Flow and Congestion Control

Flow control adjusts the packet generation rate of the source node according to the rate the receiver can handle. If the rate of incoming data is higher than a receiver can process, the receiver starts dropping them. This is a waste of resources. On the other hand, if the source node sends data packets slower than the receiver can receive, even though there are data to send and the network can convey data traffic at a higher rate, this is also a waste of resources. Flow control ensures that neither of these cases occurs.

If the data coming from multiple sources are concentrated at a link, and the rate of aggregated traffic is higher than the capacity of the link, congestion happens. In other words, congestion occurs when the traffic generation rate is higher than the network can convey for an extended period of time. Congestion control schemes aim to prevent this. The transport layer is responsible for tackling the challenges related to both end-to-end congestion and flow control. For example, TCP is responsible for congestion control in the Internet.

To control congestion and flow, TCP runs an algorithm called *slow start* based on three parameters: the receiver window, the congestion window and the threshold for congestion avoidance. The receiver window is notified by the receiver for flow control, and indicates the maximum size in kilobytes of data that the receiver can accept. The congestion window is to avoid congestion and is determined by the source node. The minimum of these two windows is the maximum size that the source can send at a time and then wait for an acknowledgement.

When a TCP connection is established, the congestion window is equal to one segment, which means the source node sends one segment. The segment size in kilobytes is equal to the maximum segment size allowed for the connection. If this segment is acknowledged before the timeout period expires, then the congestion window becomes two and the source node sends two segments. The congestion window size is doubled in each transmission compared to the last successful transmission until the receiver window size or the threshold is reached. If the threshold is reached before the receiver window size, the increase in the congestion window size becomes linear, i.e. it is incremented by one segment at each successful transmission. The threshold is 64 kilobytes at the beginning.

If a timeout occurs, i.e. the transmitted segments are not acknowledged within the expected timeout period, TCP assumes that this is due to congestion and reacts by halving the threshold and reinitiating the slow start process, as shown in the example in Figure 6.2. In this example, the maximum segment size is one kilobyte. The source sends one segment in the first transmission. When it receives the acknowledgement, it sends two segments. This exponential growth

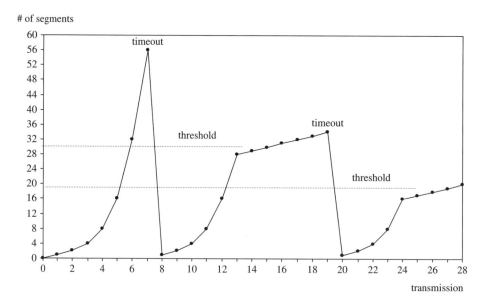

Figure 6.2   Congestion and flow control in TCP

in the number of segments continues until the receiver window size is reached (56 kilobytes for our example). When the source sends 56 segments, equal to the receiver window size, a timeout occurs due to congestion. The source halves the current window size to determine the new threshold, i.e. sets it as 28 kilobytes, and reinitiates the slow start by setting the congestion window to one kilobyte. This time the threshold is reached before the receiver window size. After the threshold, the congestion window is incremented by one segment at each successful transmission. Before reaching the receiver window size, another timeout occurs when the current window size is 34 kilobytes. The source node sets the threshold as 17 kilobytes and reruns the slow start procedure.

TCP congestion and flow control mechanisms have the following implications that make them incompatible with wireless ad hoc networks:

- TCP assumes that the communication links are very reliable, and therefore the probability that a segment loss is due to congestion is very high. This assumption is not realistic for wireless ad hoc networks where topology changes and segment losses due to node mobility and node/link failures are observed frequently.
- TCP assumes that the average end-to-end delay for connections is reasonable, and therefore the bandwidth not utilized due to slow start is negligible. This may also be in conflict with the characteristics of some wireless ad hoc networks. For example, a connection in wireless sensor networks may be through many wireless hops that make the end-to-end delay too long for TCP. Similarly, acoustic signals travel at the speed of sound, which may have an impact on the latency, making TCP impractical for underwater acoustic ad hoc networks.
- The cost of links underutilized during the slow start can be justified for tethered networks. However, the cost of wireless links is much higher than tethered links, and therefore they need to be utilized more carefully.

For these reasons, alternative versions of TCP such as TCP-Reno, TCP-Peach and I-TCP (Barakat *et al.*, 2000), which better fit wireless networks, have been developed. These later versions of TCP are generally based on the following techniques.

- **Negative acknowledgement:** in TCP the receiver acknowledges multiple segments by sending a positive acknowledgement for the last segment received. Without changing this basic scheme, negative acknowledgement for a segment not received can still be carried out by acknowledging the lost segment multiple times. When the source node receives multiple acknowledgements, i.e. three acknowledgements, for the same segment, this indicates that the receiver did not get that specific segment, and the source node resends it immediately.
- **Quick start:** right after a TCP connection is established, the source node sends one data segment and dummy segments to fill the receiver window. When the receiver acknowledges the last received segment, the source node learns the network capacity for the first round and adjusts its speed accordingly. This drastically increases the utilization of communication links with high latency.
- **Splitting:** a connection is split into multiple segments where different transport layer protocols are used. For example, a connection may contain both fiber and wireless links. While TCP is running in the fiber part of the connection, in the wireless part another transport layer protocol may be used. This solves many problems but requires demultiplexing and encapsulation of segments up to the transport layer at an intermediate node.

Although negative acknowledgements, quick start and splitting techniques may satisfy the requirements of many wireless networks, they do not completely resolve some of the constraints, such as power consumption and scalability, of wireless ad hoc networks. This is especially true for wireless sensor networks where the data generation rate is also both spatially and temporally correlated, which increases the probability of congestion.

Two challenges related to congestion control are noteworthy and require further attention for wireless sensor networks: the detection of congestion and tackling congestion (Karl and Willig, 2005). Buffer or channel statistics can be used for congestion detection. If the occupation level of a buffer at an intermediate node is over a certain level and the buffer fullness level is tending to increase, that may be a good indicator of congestion. Alternatively, a link can be sampled for a number of times during a certain time period. If the ratio between the number of times that the link is busy and the total number of samples is over a threshold value, that may also indicate congestion

To tackle congestion, either the traffic generation rate may be reduced or less important packets may be dropped. In the rate-control option, a node that detects the congestion can notify the previous nodes to reduce the number of packets they send. If this causes congestion in the previous nodes, they also notify their gradients. This may continue up to the source nodes. In the packet-dropping option, the source nodes can label their packets according to their importance, and a node that detects the congestion drops the less important packets until the congestion disappears. Both of these techniques can inspire further denial-of-service attacks. For example, a malicious node may send its neighbors false congestion reports.

6.3 Review Questions

6.1 What are the main differences between the conventional end-to-end reliability concepts and reliable end-to-end event delivery?

6.2 What are the advantages and disadvantages of enforced acknowledgement compared to selective acknowledgement?

6.3 A TCP connection is established through 10 wireless hops in an acoustic underwater ad hoc network. Assume that each hop is 100 meters on average, acoustic signals travel 1500 meters per second, 10 kilobits of data can be sent per second over a link, the source node has one megabyte of data to send, the receiver window size is 64 kilobytes, the maximum segment size is 1 kilobyte and the average processing delay for routing at each node is 1 millisecond. Assume also that the following segments do not reach the destination: 67, 112, 137. What would be the average utilization of the links for this connection if there was no other communication through them?

6.4 Is buffer status or link statistics more reliable for indicating congestion? Discuss and justify your answer.

6.5 What are the advantages and disadvantages of packet dropping compared to the rate-control technique for congestion handling?

7

Other Challenges and Security Aspects

Wireless communications create further challenges that impact on networking protocols and algorithms in any layer. These additional issues arise mainly because of factors such as mobility and scalability. Localization, synchronization, addressing, data aggregation and querying, coverage, mobility and resource management are all additional challenges faced by untethered communications and we will review these topics in this chapter.

7.1 Localization and Positioning

Localization (Bulusu *et al.*, 2000; Doherty *et al.*, 2001; Savvides *et al.*, 2001; Erdogan *et al.*, 2003; Niculescu and Nath, 2003; Patwari *et al.*, 2003) is one of the key issues in ad hoc networks, and especially in sensor networks because sensed data is almost meaningless without associating it with location data in many applications. Moreover, localization may be needed for some tasks such as spatial addressing of nodes and geographical routing. Like many other aspects related to ad hoc networks, the required level of localization accuracy differs from one application to another. For example, it may be enough to indicate in which room the measurement is made in one application. On the other hand, an accuracy level down to centimeters may be required in another application.

The first option for node localization is the global positioning system (GPS). However, GPS is not always a viable option for ad hoc networks. Nodes may be located in places where signals coming from satellites are not received with the required strength. In addition to this, GPS modules may be too expensive to attach to every node in some applications. Therefore, indirect techniques are important, especially for sensor networks (Figure 7.1).

Indirect techniques can be absolute or range-free. *Absolute* techniques can give the absolute position of a node, either according to a local or global reference coordinate system. They are generally based on triangulation, trilateration or multilateration (Figure 7.2). *Triangulation* is achieved by using the estimated angle from beacon nodes. When triangulation is used, the intersection point of the lines drawn from beacon nodes at the estimated directions gives the location of the node, as shown in Figure 7.2(a). *Trilateration* and *multilateration* are based

---

Security in Wireless Ad Hoc and Sensor Networks   Erdal Çayırcı and Chunming Rong
© 2009 John Wiley & Sons, Ltd

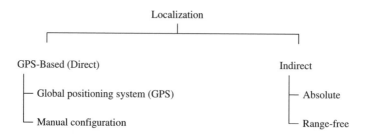

Figure 7.1   Classification of localization schemes

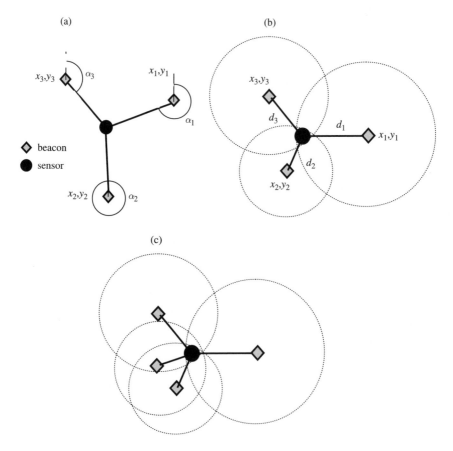

Figure 7.2   (a) Triangulation; (b) trilateration; and (c) multilateration

on the distance from the beacons. The intersection of the circles that have the related beacon
nodes at their center and radius equal to the distances from the beacon nodes is the estimated
location of the node, as shown in Figure 7.2(b) and 7.2(c).

The basic idea in multilateration is to have at least n equations to estimate n variables. For
example, when the locations (x_i, y_i) of three beacons and the distance (d_i) from a point to them

are known, the following three equations can be derived, which are enough to find out the x and y coordinates of the point:

$$(x - x_1)^2 + (y - y_1)^2 = d_1^2$$
$$(x - x_2)^2 + (y - y_2)^2 = d_2^2 \quad\quad (7.1)$$
$$(x - x_3)^2 + (y - y_3)^2 = d_3^2$$

Note that Equation (7.1) assumes that the locations of beacons and the distances to them are known accurately. This may not be the case because there can be errors in estimating the distance to the beacon nodes. One approach to tackling these errors is minimum mean square estimation (MMSE), which is based on minimizing the mean square error ε. For example, assuming that we have m beacon nodes, and we know their locations (x_i, y_i), we can estimate the distance d_i to each of them. Minimizing the mean square error ε formulated in Equation (7.2) gives us the estimated location (x, y).

$$\varepsilon = \frac{\sum_{i=1}^{m} \left(d_i - \sqrt{(x - x_i)^2 + (y - y_i)^2} \right)^2}{m} \quad\quad (7.2)$$

The distance from a beacon node can be estimated by using one of the following techniques: *received signal strength* (RSS), *time of arrival* (TOA) or *time difference of arrival* (TDOA). The technique for estimating the direction of a beacon node is called the *angle of arrival* (AOA). All these techniques have pros and cons. In RSS a node knows the location of the beacons and the strength of the signals transmitted by them. Then it estimates the distance of the beacons by using a propagation model and received signal strengths. The results may not be highly accurate due to multipath effects, other impairments such as shadowing and scattering and non line-of-sight conditions.

In TOA the node is time synchronized with the beacon nodes. It knows the location of the beacons together with the transmission time of the signals. When the node also knows the reception time, it is a simple computation to find out the distance of the beacons based on the propagation speed of the signal. The results obtained by TOA may also be impaired due to multipath effects and non line-of-sight conditions. Moreover, the propagation speed of the RF signals is too high for sensor networks where the distances between nodes are limited to only a few meters in most cases. Therefore, ultrasound signals that have lower propagation speed are used together with RF signals in TDOA. The destination receives the ultrasound and RF signals at different times although they are transmitted at the same time. Knowing the difference between the speeds of these two signals, the difference between the arrival times of them gives the distance to the source.

The AOA technique requires the usage of special antenna configurations. It may also be inaccurate due to multipath effects, non line-of-sight conditions and other sources of impairment in the wireless medium.

There are three approaches to carrying out node localization computations in ad hoc networks: *centralized*, *distributed* and *locally centralized*. In the centralized approach, all measurements by nodes are sent to a central node. The central node finds out the locations of the nodes by using these measurements, then disseminates the results. Since sensor nodes have limited computational power and memory space, this may be a viable option for some sensor network applications. Moreover, in some applications sensor nodes may not need localization information but the central node that carries out some tasks such as route optimization, optimal sensor field coverage computations, spatial data aggregation, etc., may need localization data. In addition, a centralized approach may perform better for collaborative multilateration (explained below). In the distributed approach nodes find out their locations themselves. Clusters where a central node for each cluster computes the locations of the nodes in the cluster are established in the locally centralized approach.

In collaborative multilateration, sensor nodes collaborate with the other sensor nodes for localization when they do not receive signals from enough beacons. For example, two sensor nodes that receive signals from two beacons can collaborate to alleviate the lack of signal from the third beacon, as shown in Figure 7.3. The basic idea is again to have at least n equations to estimate n variables. Details on collaborative multilateration can be found in Savvides *et al.* (2001).

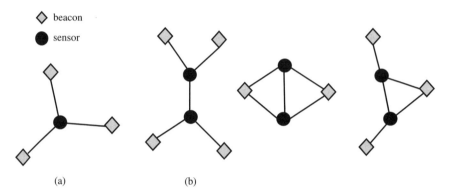

Figure 7.3 Collaborative multilateration (a) one-hop multilateration; (b) two-hop collaborative multilateration

Indirect techniques can also be range-free. Two examples of range-free techniques are sectoral sweepers (SS) (Erdogan *et al.*, 2003) and centroid schemes, as shown in Figure 7.4. Although the resolution of the range-free techniques is typically not as high as the other techniques explained in this section, they are simple enough to implement without any additional hardware or software components at the nodes. Moreover, the resolution of them is high enough for many ad hoc and sensor network applications.

The SS scheme is based on task dissemination by using directional antennae. Each task is also associated with minimum and maximum RSS values and unique task identification. When a sensor node reports for a task, the task identification implies a specific region, indicated by the direction of the antenna, that disseminates the task and minimum/maximum RSS values specified for the task. Note that the borders of the task region cannot be very well defined, but are instead a little amorphous, as shown in Figure 7.4(a), due to multipath and non line-of-sight effects.

Creating overlapping task regions for the same task can enhance the resolution of the SS scheme. When a sensor node reports for multiple tasks, the intersecting area of the reported

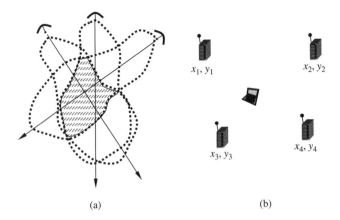

(a) (b)

Figure 7.4 Range-free techniques (a) sectoral sweepers; (b) centroid scheme

task regions is the location of the node. This intersection area is smaller than a task region. Hence, location estimation with higher resolution can be achieved.

In the centroid technique nodes calculate the centroid of a set of anchor nodes. Equation (7.3) gives the centroid of beacons when the locations of beacons are known and the received signal strengths from them are close to each other.

$$(x, \ y) = \left(\frac{\sum_{i=1}^{n} x_i}{n}, \frac{\sum_{i=1}^{n} y_i}{n} \right) \tag{7.3}$$

Node localization techniques also present new opportunities for attacks. For example, a malicious node may introduce itself as a beacon and disseminate false location information, which hinders node localization. There are many other security attacks against node localization schemes and these are examined in more detail in Chapter 8.

7.2 Time Synchronization

Time synchronization is also important for ad hoc networks and especially sensor networks, not only because of the requirements of the protocols in various layers such as medium access control (i.e. scheduling) and the network layer (i.e. routing and aggregation), but also sensed data often need to be related to a time. The factors influencing time synchronization in large systems are as follows (Mills, 1994, 1998; Levine, 1999; Elson *et al.*, 2002):

- **Temperature:** temperature variations during the day may cause the clock to speed up or slow down (a few microseconds per day).
- **Phase noise:** access fluctuation can occur at the hardware interface; response variation of the operating system to interrupts, jitter in delay, etc.
- **Frequency noise:** the frequency spectrum of a crystal has large sidebands on adjacent frequencies.

- **Asymmetric delay:** the delay of a path may be different for each direction.
- **Clock glitches:** hardware or software anomalies may cause sudden jumps in time.

These factors contribute to the differences between the times given by the clocks of two nodes. These differences in time can be classified into the following categories (Ganeriwal *et al.*, 2005):

- **Offset (o):** nodes may be started at different times. Therefore, node A may have a clock C_A different from the clock C_B of node B when the network starts at time t_0.

$$o = C_A(t_0) - C_B(t_0) \qquad (7.4)$$

- **Skew (s):** factors like frequency noise and hardware may make the crystals of nodes run at different frequencies. This causes clock skew, which may be ±30–40 parts per million (ppm) for sensor node hardware. Skew may provide instances where two nodes get closer or further away based on the offset. The skew-related change per unit time t is constant.

$$s = \frac{\partial C_A}{\partial t} - \frac{\partial C_B}{\partial t} \qquad (7.5)$$

- **Drift (d):** factors like temperature, phase, asymmetric delay and clock glitches may change the offset between two nodes over time. Since these factors are temporally variable, the change in the clocks, called *drift*, per unit time is not a fixed value.

$$d = \frac{\partial^2 C_A}{\partial t^2} - \frac{\partial^2 C_B}{\partial t^2} \qquad (7.6)$$

Time synchronization algorithms for ad hoc networks can be categorized according to three criteria: distribution of the synchronization procedures, roles of the nodes in synchronization and the accuracy in synchronization, as shown in Figure 7.5. In centralized time synchronization, nodes are synchronized to a central time server. The network time protocol (NTP) (Mills, 1994) falls into this category. NTP is the protocol that has been adopted most widely in the Internet. In NTP there are time servers synchronized by external time sources, such as GPS. All the other nodes in the network get synchronized with these time servers. In the distributed

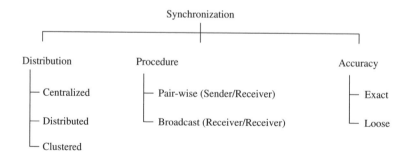

Figure 7.5 Time synchronization schemes

approach, nodes do not synchronize with a central time server but with the nodes that they need to get synchronized. The reference broadcast synchronizations (RBS) scheme (Elson *et al.*, 2002) is an example of the distributed approach where nodes get synchronized with the other nodes in the network by using a reference time stamp from a central node. In the third approach, a network self organizes for time synchronization. Nodes are synchronized within clusters, and then clusters are synchronized with each other.

We can also categorize the synchronization schemes according to the flow of the synchronization procedure and the tasks of various nodes in this flow. There are two classes in this respect: sender/receiver synchronization is basically for pairwise synchronization and receiver/receiver synchronization is for broadcast synchronization.

In sender/receiver synchronization the sender sends a time-stamped synchronization message and the receiver synchronizes itself with the sender according to the time stamp in the message. The timing-sync protocol for sensor networks (TPSN) (Ganeriwal *et al.*, 2003) is a sender/receiver algorithm.

In receiver/receiver synchronization, one node broadcasts periodically a synchronization message, which is time stamped. This message is not used by the other nodes in the network to synchronize with the sender, but the time stamp in that message constitutes a reference time for all the nodes receiving the message. Later, when a node needs to synchronize with the other nodes receiving the same synchronization message, they exchange their offset from the time stamp in the message. One advantage of this scheme is that the nodes learn about the offset and drift of the other nodes. RBS falls into this category.

Finally, nodes may not need to synchronize exactly with each other. Sometimes it may be too costly to keep all the nodes exactly synchronized with each other and clock offsets up to a certain threshold can be tolerated. Time synchronization protocols that do not guarantee exact time synchronization but clock offsets below a certain threshold are called *loose synchronization schemes*.

7.3 Addressing

The unique features and application requirements of wireless ad hoc networks present new challenges in node addressing. Fixed and universal addressing of nodes is not a viable option for many ad hoc and sensor network applications. Therefore, attribute-based naming or local identification of nodes have thus far been considered to address a specific node for various purposes such as node management, data querying, data aggregation or routing. We categorize the node-addressing techniques applicable to ad hoc and sensor networks as follows:

- **Attribute-based naming and data-centric routing:** attribute-based naming (Intanagonwiwat *et al.*, 2000) is one of the earliest techniques used for node addressing in wireless sensor networks. In this technique, nodes that measure certain values for a specified attribute are called, for example, 'nodes that measure more than 35°C temperature.'
- **Spatial addressing:** spatial addressing is especially useful in applications such as intrusion detection and target tracking where queries are mainly based on node locations. It is also needed for spatial data aggregation and geographic routing schemes. In this technique the borders of a region are defined, and then the nodes inside, outside or in a buffer zone that have a certain depth along this border are queried. Borders of the region can be detected by

an event boundary detection algorithm (Ding *et al.*, 2005) or can be specified by giving a series of geographical locations.

- **Sectoral sweepers** (Erdogan *et al.*, 2003): spatial regions can also be specified by using a directional antenna, where sensor nodes that receive a signal with a certain range of received signal strength indicator respond to a query.
- **Using local identification for nodes and mapping the destination of the user queries to the local identifications:** in this scheme, every node is addressed by a local identification in the sensor field. The destination of incoming task packets or queries is mapped to these local identifications by the intermediate nodes. Users may indicate a destination by using either an attribute-based naming or spatial addressing scheme. A gateway node maps this address to a local identification.
- **Address reuse:** address reuse is especially useful for MAC layer addressing. The same addresses can be assigned to multiple nodes as long as this does not cause a conflict. A distributed protocol for address reuse is proposed in (Schurgers *et al.*, 2002). In this protocol, a node first broadcasts a 'Hello' message. The nodes that hear this message reply with an 'Info' message where they declare their local addresses. The node that broadcasts the 'Hello' message randomly selects an address not used by its neighbors by using the data in the 'Info' messages. If there is a conflict due to the hidden node problem, the first node that detects this conflict sends a 'Conflict' message to both of the nodes, and one of them changes its address.

Address reuse, local identifications and spatial addressing techniques create new challenges for authentication.

7.4 Data Aggregation and Fusion

In sensor networks data aggregation is important for two reasons: first, it is needed to fuse data and to derive information from data; second, it is perceived as a means of reducing the communications overhead. Therefore, there has been considerable research effort on data aggregation, which can be classified as follows:

- **Temporal or spatial aggregation:** data may be aggregated in a time-based or location-based fashion. For example, the temperature readings for every hour or temperature readings for various regions in a sensor field may be averaged. A hybrid approach – a combination of time- and location-based aggregation – may also be used.
- **Snapshot or periodical aggregation:** data aggregation may be made in a snapshot, i.e. one time, on the receipt of a query. Alternatively, temporally aggregated data may be reported periodically.
- **Centralized or distributed aggregation:** a central node can gather and then aggregate data or data may be aggregated while being conveyed through a sensor network. A hybrid approach is also possible, where clusters are set and a node in each cluster aggregates the data from the cluster.
- **Early or late aggregation:** data may be aggregated at the earliest opportunity, or aggregation of data may not be allowed before a certain number of hops in order not to hinder collaboration among the neighboring nodes.

Data correlation, data association and fusion techniques are used to correlate the data coming from multiple sensors to an event, associate the data coming from sensors with multiple events and fuse the data into a common picture. Data aggregation schemes should not hinder these tasks.

7.5 Data Querying

One of the most challenging tasks in sensor networks is to synthesize the information requested by users from the available data measured or sensed by a large number of sensor nodes. Since there are a vast number of nodes with stringent energy constraints in a sensor network, it may not be feasible to gather every node reading for central processing. Instead, effective data querying and aggregation techniques are needed. In this section we focus on data querying in sensor networks.

Data queries in sensor networks may be continuous and periodical, continuous and event-driven or snapshot, i.e. one-time, queries. We can also categorize sensor network queries into aggregated or nonaggregated. Queries can also be complex or simple. Finally, queries for replicated data can be made. Users should be able to make any of these types of queries by using the data-querying scheme for sensor networks. One approach to realizing this is to perceive a sensor network as a distributed database (Figure 7.6), such as the approach in data aggregation and dilution by modulus addressing (DADMA) (Cayirci, 2003).

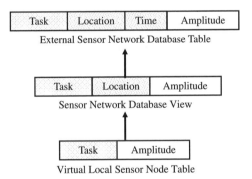

Figure 7.6   Sensor network as a distributed database

7.5.1 Database Approach

In DADMA a sensor network is perceived as a distributed relational database composed of a single view that joins local tables located at sensor nodes. Records in local tables are the measurements made upon a query arrival and consist of two fields, namely *task* and *amplitude*. Since a node may have more than one sensor attached to it, the *task* field, e.g. temperature, humidity, etc., indicates the sensor that makes the measurement. Nodes have limited memory capacity and they do not store the results of every measurement. Therefore, the *task* field is the key field in the local tables created upon a query arrival. This perception of sensor networks makes relational algebra practical for retrieving the sensed data without much memory requirement.

A sensor network database view (SNDV) can be created temporarily either at a central node or at an external proxy server. An SNDV record has three fields: *location, task* and *amplitude*. While data are being retrieved from a node, the sensed data are joined by the *location* of the node. Since multiple nodes may have the same type of sensor, i.e. multiple sensors can carry out the same sensing *task*, the *location* and *task* fields become key in an SNDV. For many WSN applications, the sensed data need to be associated with location data. For example, in target tracking and intrusion detection sensor networks, sensed data are almost meaningless without relating them to a location. Therefore, location awareness of nodes is a requirement imposed by many WSN applications. If the location data are not available and not important to the application, the local identification field for the node replaces the location field.

It is also possible to maintain a database in a remote proxy server where the records obtained from queries, i.e. the records at an SNDV, are stored after being joined with a *time* label. For example, a daemon can generate queries at specific time intervals and insert the records in the SNDV resulting from these queries into the database after joining them with a *time* field. Note that each query results in a new SNDV where the results of the query are gathered temporarily.

In DADMA (Cayirci, 2003), a query is started by a statement that has the structure given below. Note that the standard SQL notation is used in this statement except for the last field starting with the *'based on'* keyword.

Select [task, time, location, [distinct | all], amplitude,

[[avg | min |max | count | sum] (amplitude)]]

from [any , every , aggregate m , dilute m]

where [power available [<|>] *PA* |

location [in | not in] *RECT* |

t_{min} < time < t_{max}|

task = t |

amplitude [<|==|>] a]

group by task

based on [time limit = l_t| packet limit = l_p| resolution = r | region = xy]

A user can retrieve a subset of data fields available in an SNDV and can aggregate *amplitude* data either by grouping data based on task and/or by using the *aggregate m* function given in Equation (7.7). Some of the nodes may also be excluded from a query by the *dilute m* function given in Equation (7.8).

$$f(x) = x \text{ div } m \tag{7.7}$$

$$f(x) = (x/r)\text{mod } (m/r) \tag{7.8}$$

where

x is the grid location of a node relative to one of the axes,
r is the resolution in meters, and
m is the dilution or aggregation factor.

When the *dilute m* command is given by the user, every node first uses Equation (7.8) to find out its location indices on the horizontal and vertical axes and then compares these indices with the region values x and y sent in the *'based on'* field of the query. If they match, the node replies to the query. For example, the location indices of a node at location {46, 74} are {3, 1} for $m = 8$ and $r = 2$. Therefore, if the region value in the query is {3, 1}, this sensor should respond. Hence, only the snodes in $r \times r$ meter squares located in every m meters react to the query while the others stay idle. This is a practical technique, especially when nodes are deployed randomly according to the uniform distribution, and the sensor network is monitoring environmental conditions such as temperature, humidity and pressure.

For the same example, the indices found by using Equation (7.7) are {5, 9}. When the *aggregate m* command is received, the values measured by a node are aggregated with the values measured by the other nodes having the same indices. Hence, we can address the snodes at certain geographic locations and aggregate data based on the location of nodes.

7.5.2 Task Sets

The idea of task sets is to divide a sensor field into subregions and assign a specified number of nodes in each subregion to every task set (Cayirci *et al.*, 2006a). The number of nodes in each subregion varies because of the nonhomogeneous distribution of nodes. Hence, the cost of querying the sensor field varies in different subregions. To balance this cost, forming task sets (TSs) with a specific number of nodes in each quadrant is proposed in Cayirci *et al.* (2006a). Using TSs, users have an initiative by which to trade off accuracy/reliability and communications cost. The number of nodes in a task set indicates the resolution of the data that can be collected by querying the task set. A higher number of nodes in a task set implies higher accuracy and reliability. On the other hand, more power is consumed as the number of nodes in a task set increases.

For example, TS1 may be specified as the two nodes that have the highest power available in every subregion. Similarly, TS2 may be specified as all nodes that are not in TS1. After the task sets are formed, the queries are sent either to TS1 or TS2.

7.5.3 Other Data-Querying Schemes

In the active query forwarding in sensor networks (ACQUIRE) scheme (Sadagopan *et al.*, 2003), each node that forwards a query tries to resolve it. If a node resolves the query, it does not repeat it but sends the result back. Nodes collaborate with their n-hop neighbors to resolve a query. The parameter n is called the *look-ahead parameter*. If a node cannot resolve a query after collaborating with its n-hop neighbors, it forwards it to another neighbor. When the look-ahead parameter n is 1, ACQUIRE performs flooding in the worst case.

Mobility-assisted resolution of queries in large-scale mobile sensor networks (MARQ) (Helmy, 2003) makes use of the mobile snodes to collect data from the sensor network.

In MARQ, every node has contacts that are some of the other nodes. When contacts move around, they interact with other nodes and collect data. Nodes collaborate with their contacts to resolve the queries.

The sensor query and tasking language (SQTL) (Shen *et al.*, 2001) has been proposed as an application layer protocol that provides a scripting language. SQTL supports three types of event, defined by the keywords receive, every and expire. The *receive* keyword defines events generated by a node when the node receives a message; the *every* keyword defines events that occur periodically due to a timer timeout; and the *expire* keyword defines events that occur when a timer has expired. If a node receives a message that is intended for it and contains a script, the node then executes the script.

7.6 Coverage

The term *coverage* indicates two things in wireless ad hoc networks: the area of communications covered by the network or the area that can be observed by the sensors in a sensor network. Providing the maximum coverage by the same amount of nodes is an important challenge related to the following factors (Cardei and Wu, 2005):

- the node deployment scheme;
- the sensing and communications range;
- energy efficiency and connectivity requirements;
- the algorithm paradigm, i.e. centralized or distributed.

The coverage problem can be categorized into three classes (Cardei and Wu, 2005):

- In *area coverage* the objective is to cover an area, which, for the sensing coverage problem, means ensuring that every point in a given area can be observed, and, for the communications coverage problem, means that a node at any point in the area can access the network.
- In *point coverage* the objective is to ensure that a given set of points is covered by the network.
- In *barrier coverage* the objective is to ensure that there is no hidden path through the network, i.e. an intruder cannot go through the network without crossing the coverage area of at least one node.

Two approaches are generally followed in solving coverage problems. In the first approach, the nodes are assumed to be deployed randomly according to a distribution, and the minimum number of nodes that satisfies a given probability of coverage is determined. In the second approach, it is assumed that the nodes can be deployed at certain locations, and the location for each node is determined such that the maximum coverage for the given number of nodes can be achieved. Various algorithms that follow one of these approaches can be found in (Cardei and Wu, 2005; Cayirci *et al.*, 2006b).

New attacks can be designed to discover uncovered areas or to create intrusion tunnels through a sensor network by compromising the nodes that run coverage-related algorithms.

7.7 Mobility Management

Mobility management is required in networks where the nodes are mobile. This is a particularly challenging issue in infrastructured networks, although the infrastructure is not mobile because they are designed to support seamless global roaming. In ad hoc networks mobility is generally limited within the network. Therefore, mobility management attracts fewer researchers for ad hoc networks. In infrastructured, i.e. cellular, networks mobility management has two tasks: location management and handoff management. Location management is to keep track of users so that calls arriving for them can be directed to their current location. Handoff management enables users to roam while they have an active connection. A mobile node that is communicating can leave the coverage area of an access point and enter another one, and handoff management allows this without terminating the already established connection. Since there are no access points but routes made up of multiple wireless hops in ad hoc networks, routing protocols tackle the route changes during an active connection. Location management can also be an issue for ad hoc networks because nodes can roam among several ad hoc networks. Therefore, location management techniques similar to that in mobile IP are also needed for ad hoc networks.

Location management in cellular networks involves two tasks: *location update* and *paging*. The location information about a mobile terminal (MT) is maintained by location updates. In current systems, cells, i.e. access points, are grouped into location areas (LAs). An LA may have one or multiple cells assigned to it. An MT reports its location whenever it enters a new LA. Since an LA consists of a number of cells, the exact location of an MT should be determined for a call delivery. This is done by paging the cells in the last registered LA.

When an LA is comprised of a group of cells that are permanently assigned to that LA, and is fixed for all MTs, the location management scheme is called *static*. *Dynamic* location management techniques are more adaptive to the mobility characteristics of MTs. They allow dynamic selection of the location update parameters and reduce the signaling traffic due to location management. *Time-based, movement-based and distance-based location update techniques* are well-known dynamic schemes (Cayirci and Akyildiz, 2002; 2003). In the *time-based* technique, an MT performs location updates periodically at a predefined time interval. In the *movement-based* technique, location updates are initialized after crossing a certain number of cell boundaries. The *distance-based* location update is performed when the distance from the last registered cell exceeds a predefined value.

There are also other dynamic location update techniques in the literature. In the *direction-based* scheme, an MT reports its location only when its movement direction changes. In the *selective* location update technique, location updates are not performed in every LA, and some LAs are skipped based on the transition probabilities and the cell dwell times. The *state-based* location update technique, where an MT decides to update its location based on its current state, is another dynamic technique. Policies that are combinations of the known techniques, such as time and distance based schemes, are also proposed in the literature (Cayirci and Akyildiz, 2002).

The primary goal of location updates is to reduce the paging cost, and the performance of the paging strategies are closely related to the location update scheme. The location update schemes must provide the network with enough information such that the paging cost is reduced under a given delay requirement. There is a tradeoff between the paging cost and the paging delay as well as between the paging and update costs. As the resolution of the updated location information increases, the number of cells to be paged decreases. Similarly, the longer the paging delays, the lower the number of paged cells becomes.

The least paging delay is guaranteed by the *blanket polling* technique where all cells in an LA are simultaneously paged upon a call arrival. This scheme has high paging cost. *Selective paging* is an alternative to *blanket polling* and it reduces the paging cost but increases the paging delay. In selective paging, the location of the MT is predicted based on its location probability and cells are paged sequentially, starting from the cells where the MT is most likely to be present. Several other paging strategies such as the *shortest distance first* and *velocity paging* have also been proposed. In velocity paging, the system calculates the maximum distance that a mobile can travel based on its average velocity and the last registration time, and pages the cells that are within this distance from the last registered cell. Paging schemes with delay bounds as a QOS constraint have also been introduced, where the location areas are partitioned into clusters. The number of clusters in a location area guarantees that the system can page them sequentially without exceeding the given delay bound.

7.8 Cross-layer Design

We would like to end Part One of the book by dedicating one short section to the cross-layer approach. Our goal in this section is just to define it and make the reader think about the security implications of cross-layer protocols. We do not provide details about a set of protocols that fall into this category. That is beyond the scope of this book.

The layered approach of OSI is designed to provide interoperability and reusability for the networking protocols and schemes. However, it does not always lead to the optimum solutions. In the OSI layered approach, transparency from the lower layer details, interoperability and reusability of protocols are achieved at the expense of a more costly protocol stack compared to the cross-layer design.

Since WASMs introduce some very stringent constraints, cross-layer protocols are common in our domain. Cross-layer optimization allows communication between layers, which results in network performance improvements. Cross-layer interactions, when used properly, can boost efficiency in ad hoc networks by passing information among the layers, which can lead to effective adaptation to the dynamic environment. For example, the network layer can interact with the data link layer, enabling the links where higher contention is observed to be avoided in route selections, and an adaptive cross-layer congestion control scheme that better fits the requirements of WASMs can be designed. Similarly, medium access schedules and routing decisions can be optimized jointly, leading to higher efficiency and better QOS.

There are two approaches to cross-layer design (Aune, 2004): *evolutionary* and *revolutionary*. An evolutionary approach tries to extend the layered structure in order to maintain compatibility. Most cross-layer designs are evolutionary. Simple solutions that allow two or three layers to share information can improve efficiency and adaptability considerably. Revolutionary approaches do not seek to extend an existing architecture or to maintain compatibility. Although they are not common, there may be revolutionary cross-layer designs that offer better efficiency and adaptability than other approaches, and therefore that may be very useful for specific architectures.

Cross-layer designs can create opportunities for new kinds of security attacks, which we examine in detail in the next chapter. For example, an adversary may exploit a routing scheme adaptive to link characteristics by degrading the characteristics of certain links to promote another link that leads to a malicious node. On the other hand a security scheme can be optimized for multiple layers to reduce the security overhead, or a security scheme at a certain layer can be designed cross-layer such that it may be adaptive to the information coming from the other layers. Please keep this short section about cross-layer approaches in your mind when you are reading Chapter 8 in particular.

7.9 Review Questions

7.1 List and explain the techniques that can be used to find out the distance to a beacon node. What are the weaknesses of these techniques?

7.2 A node receives signals from three anchor nodes at coordinates (100, 220), (150, 180) and (60, 80). Note that the coordinates indicate the distance in meters from a reference point.

(a) Estimate the node's location by using the centroid scheme.

(b) Assume that the nodes also send ultrasonic signals that have perfect transmission time synchronization with RF signals, and ultrasonic signals travel at the speed of sound while RF signals travel at the speed of light. What are the absolute coordinates of the node if the time differences of arrival for ultrasonic and RF signals are as follows for each anchor node?

    anchor node 1:   67 ms
    anchor node 2:   100 ms
    anchor node 3:   80 ms

(c) Can you find out the location according to the values in (b)? If you cannot, what could be the reason for that?

7.3 What are the reasons for clock skew? Examine technical specifications of at least one microprocessor for the clock skew value.

7.4 Is it possible to use globally unique addresses for sensor nodes? If not, what are the reasons? Explain the alternative approaches to addressing sensor nodes.

7.5 When is late aggregation used?

7.6 What are the differences between data association, data correlation and data fusion?

7.7 Do you think that it is a viable approach to perceive a sensor network as a database to query? Propose a technique to do so.

7.8 What is the difference between area coverage and barrier coverage? Which problem is more challenging? Why?

7.9 Give two tasks in location management. Briefly describe them.

Part Two

Security in Wireless Ad Hoc, Sensor and Mesh Networking

8

Security Attacks in Ad Hoc, Sensor and Mesh Networks

In this chapter we provide taxonomy for both security attacks and attackers. First, security attacks are categorized and examples of attack scenarios are given. Different types of attacker with various motives can carry out the same type of attack. Defense mechanisms may need to be sensitive not only to the type of attack but also the type of attacker. Therefore, attackers are also classified and their motives are explained at the end of the chapter.

8.1 Security Attacks

Security attacks can be categorized into two broad classes: *passive* and *active* attacks. Passive attacks, where adversaries do not make any emissions, are mainly against data confidentiality. In active attacks, malicious acts are carried out not only against data confidentiality but also data integrity. Active attacks can also aim for unauthorized access and usage of the resources or the disturbance of an opponent's communications. An active attacker makes an emission or action that can be detected.

Apart from security attacks, needlessness is also an important security threat. By mistake, users can expose nodes to threats like tampering and destruction, and classified data and resources to unauthorized access. Security and fault-tolerance schemes should also tackle the security and safety challenges created by careless use or unpredicted events.

8.1.1 Passive Attacks

Before we examine passive attacks, please note that RF is not the only wireless medium. There are wireless carriers, such as infrared and other optical channels, which are more resilient to tapping. These kinds of channel are directed and transmissions to them are spatially limited. To tap them, the receiver of the adversary needs to be located accordingly, which makes the task of the adversary more difficult and the probability that the adversary is detected higher.

Security in Wireless Ad Hoc and Sensor Networks   Erdal Çayırcı and Chunming Rong
© 2009 John Wiley & Sons, Ltd

In passive attacks attackers are typically camouflaged, i.e. hidden, and tap the communication lines to collect data. Passive attacks can be grouped into eavesdropping and traffic analysis types (Figure 8.1).

Figure 8.1 Passive attacks

8.1.1.1 Eavesdropping

Classified data can be *eavesdropped* by tapping communication lines, and wireless links are easier to tap. Therefore, wireless networks are more susceptible to passive attacks. In particular when known standards are used and plain data, i.e. not encrypted, are sent wirelessly, an adversary can easily receive and read the data and listen to or watch audio–visual transmissions. For example, an adversary can easily eavesdrop credit card numbers and passwords when they are transmitted plainly over unsecured wireless links.

Actually, ad hoc and sensor networks are a little more secure against tapping compared to other longer range wireless technologies because signals are sent over shorter distances. An adversary needs to get close enough to the attacked node to be able to tap. If the facility where these wireless technologies are used has enough space controlled against intruders, i.e. unauthorized people and devices, it becomes more secure. However, they can never become as secure as tethered communications. An insider close enough to the target terminal can receive all frames to or from it, store them in some medium and take them out of the facility. These risks can be reduced by careful inspections of everyone leaving or by monitoring the electromagnetic emissions from the facility. Still, the risks are much higher when wireless technologies are used.

Moreover, the existence of wireless communications makes the implementation of multiple networks with different security levels at a single facility much more difficult. For example, if there are both classified networks and a network attached to the Internet in the same facility and wireless access to the classified networks is allowed, the decoupling of the Internet and the classified networks can become very difficult due to passive attacks and needlessness. Note that not allowing untethered communications does not make the security risks disappear, but allowing them increases the risks. Needlessness, insiders and emanation security are always issues whether wireless communications are allowed or not.

We would like to differentiate *privacy* from *confidentiality* in this section. Sentient environments and ubiquitous computing enabled by wireless ad hoc and sensor networks may be abused in order to access not only confidential data but also private information. For example, the cameras for a security system may be attacked passively to observe the private lives of others. Analysis of unclassified data may also lead to private information. Therefore, some systems and data not considered confidential at first glance may be private and should be protected.

For privacy protection, anonymity is important. Attacks against privacy may start with attacks against anonymity. An adversary first needs to know which node serves which individual and for what purpose. Similarly, the adversary needs to know which data packet is coming from which node. After this is achieved, the collected data may become more meaningful. Therefore, privacy and confidentiality can be enhanced by anonymity.

8.1.1.2 Traffic Analysis

As well as the content of data packets, the traffic pattern may also be very valuable for adversaries. For example, important information about the networking topology can be derived by analyzing traffic patterns. In ad hoc networks, especially in sensor networks, the nodes closer to the base station, i.e. the sink, make more transmissions than the other nodes because they relay more packets than the nodes farther from the base station. Similarly, clustering is an important tool for scalability in ad hoc networks and cluster heads are busier than the other nodes in the network. Detection of the base station, the nodes close to it or cluster heads may be very useful for adversaries because a denial-of-service attack against these nodes or eavesdropping the packets destined for them may have a greater impact. By *analyzing the traffic*, this kind of valuable information can be derived.

Traffic analysis can also be used to organize attacks against anonymity. Detecting the source nodes for certain data packets may also be a target for adversaries. This information helps to detect the location of events, weaknesses, capabilities and the functions or the owners of the nodes.

Moreover, traffic patterns can pertain to other confidential information such as actions and intentions. In tactical communications, silence may indicate preparation for an attack, a tactical move or infiltration. Similarly, a sudden increase in the traffic rate may indicate the start of a deliberate attack or raid. Similar information can also be derived by traffic analysis in civilian networks. Traffic analysis can be carried out to list the frequent contacts of every end terminal – called *friendship trees* in intelligence efforts. By sifting through the network traffic, the contacts of a node can be determined. Analyzing the signals of contacts together with the targeted node itself is often more meaningful.

One of the following techniques may be used for traffic analysis:

- **Traffic analysis at the physical layer:** in this attack only the carrier is sensed and the traffic rates are analyzed for the nodes at a location.
- **Traffic analysis in MAC and higher layers:** MAC frames and data packets can be demultiplexed and headers can be analyzed. This can reveal the routing information, topology of the network and friendship trees.
- **Traffic analysis by event correlation:** events like detection in a sensor network or transmission by an end user can be correlated with the traffic and more detailed information, e.g. routes, etc., can be derived.
- **Active traffic analysis:** traffic analysis can also be conducted as an active attack. For example, a certain number of nodes can be destroyed, which stimulates self organization in the network, and valuable data about the topology can be gathered.

8.1.2 Active Attacks

In active attacks an adversary actually affects the operations in the attacked network. This effect may be the objective of the attack and can be detected. For example, the networking services may be degraded or terminated as a result of these attacks. Sometimes the adversary tries to stay undetected, aiming to gain unauthorized access to the system resources or threatening confidentiality and/or integrity of the content of the network. We group active attacks into four classes, as shown in Figure 8.2.

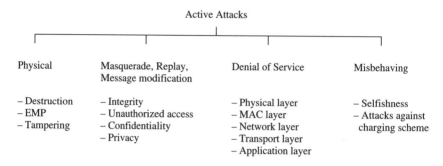

Figure 8.2   Active attacks

8.1.2.1 Physical Attacks

An adversary may *physically damage hardware* to terminate the nodes. This is a security attack that can also be considered to fall in the domain of fault tolerance, which is the ability to sustain networking functionalities without any interruption due to node failures. Physical attacks against hardware may become a serious issue, especially in tactical communications and sensor networks. Sensor nodes may be deployed unattended in regions accessible by the adversary. Therefore, they can be moved out of the sensor field or destroyed. When these risks are imminent, nodes need to be resilient to physical attacks.

When nodes are unattended and can be reached physically by the adversary, they can be attacked by *tampering* techniques, such as microprobing, laser cutting, focused ion-beam manipulation, glitch attacks and power analysis (Komerling and Kuhn, 1999). Node tampering can help in masquerading and denial-of-service attacks, which we explain in the following sections. Therefore, tamper resilience is an issue that needs to be considered carefully in many sensor network and tactical communications applications.

We can group node-tampering schemes into two classes: *invasive* tampering and *noninvasive* tampering. Invasive techniques aim to gain unlimited access to a node. In noninvasive attacks, unlimited access to the node is not the intention. Instead, by analyzing the behavior of a node, such as the power consumption, or the execution timings of the algorithms for various inputs, confidential data about the procedures and keys used by the encryption schemes can be derived.

Electromagnetic pulse (EMP) attacks are also among the threats that can be listed within physical security attacks. An EMP is a short-duration burst of high-intensity electromagnetic energy that can produce voltage surges, which can damage electronic devices within range. An EMP is a natural result of nuclear explosions. Today, portable devices that can generate EMPs are also available. Although there are still unsolved issues related to the practicability of EMP technologies, EMPs are a threat for all kinds of electrical devices in the tactical field.

Again, this can be considered as part of the fault tolerance domain. It is possible to build electronic devices that are more resilient to EMPs. Therefore, we have listed EMP attacks as a type of security attack.

8.1.2.2 Masquerade, Replay and Message Modification

A *masquerading* node acts as if it is another node. Messages can be captured and *replayed* by masquerading nodes. Finally, the content of the captured messages can be *modified* before being replayed. Various scenarios and threats can be developed based on these approaches.

Ad hoc and sensor networks introduce particular advantages for masquerading. In mobile ad hoc networks, nodes may change their location in the network. This location is not given or fixed, and self-forming and self-healing mechanisms are counted on to adapt to topology changes. Since reactive techniques are preferred for routing and topology may not be maintained, it may be difficult to check the consistency of a node's access point to the network. Moreover, it may not be possible to check if the node has already accessed another point in the network. Sensor networks are even easier in terms of masquerading because global identifications may not be used in sensor networks. Instead, techniques like data-centric routing and address reuse may be the addressing scheme.

Masquerading, message replay and content modification can be used to attack the integrity of the content of messages or services in a network. Sensor networks in particular have several network functions susceptible to special kinds of security attack because they are based on a collaborative effort of many nodes. For example, node localization schemes may be subject to one of the following security attacks:

- A malicious node may act as a beacon and disseminate its location wrongly. This hampers the node localization procedure when the node uses beacon signals transmitted by the malicious node for triangulation or multilateration.
- A beacon may be tampered with and introduce wrong location data, transmit beacon signals with less or more power than expected to impair received signal strength indicator based schemes or slightly desynchronize the transmission of RF and ultrasonic signals if the time difference of arrival algorithm is used.
- Beacon signals may be replayed by a malicious node.
- Beacon nodes may be destroyed by physical attacks.
- An obstacle may be placed between beacon nodes and the network to block the direct line of sight.

There are many more attack scenarios that may be detrimental to the node localization schemes. In addition to node localization schemes, the integrity of the following services, explained in Chapter 7, may also be subject to similar security attacks:

- In-network data aggregation and fusion make sensor networks more sensitive to replay and content modification attacks because changing the content of an aggregated message may change the data provided by many nodes.
- Time synchronization is also a vulnerable service for masquerading attacks. Several insiders that inject false time synchronization messages may prevent the system from achieving

time synchronization. Time synchronization can be especially sensitive to replay attacks. A malicious node can jam a time synchronization message at a certain part of a network, and then replay the message at that part after a very short delay. This may prevent correct time synchronization and create considerable detrimental effects on all services that rely on the accuracy of the synchronization protocol.

- Data correlation and association techniques are also impaired when node localization or time synchronization services are attacked.
- By modifying the contents of the messages, event and event boundary detection algorithms can be hindered.
- Similarly, node management systems can be hampered by modifying the messages that report node status or convey commands for node management.

An improved version of masquerading is a *sybil* attack, where a malicious node introduces itself as multiple nodes. Having multiple identifications can be very useful for a malicious node. For example, a sybil attack can be conducted against data correlation and aggregation techniques. A node that sends multiple values with different identifications can change an aggregated value considerably. A sybil attack can also threaten multiple path routing schemes, node localization, etc. Multiple identifications can also help to keep the attacks hidden, i.e. stealthy attacks.

Note that we can also consider attacks against the integrity of services as denial-of-service (DoS) attacks because they reduce the availability of some services. We explain DoS attacks in detail in the following section.

Masquerading, message replay and content modification can also be used against confidentiality by making the other nodes send the confidential data to a malicious node or by accessing the confidential data. They can also be used as techniques for gaining unauthorized access to system resources.

An adversary masquerades in *phishing*, which means deceiving someone in order to make him/her give confidential information voluntarily. The term 'phishing' is a combination of two words – password and fishing – which define this attack very well. A malicious node that impersonates an authorized node can ask another node to give information about passwords, keys, etc.

Masquerading is also one approach for preserving the anonymity of a malicious node that provides illegal or unethical content, or one that attacks or gains illegal access to a remote system, e.g. a major government or banking database.

8.1.2.3 Denial-of-Service Attacks

A denial-of-service (DoS) attack mainly targets the *availability* of network services. A DoS is defined as any event that diminishes a network's capacity to perform its expected function correctly or in a timely manner. A DoS attack is characterized by the following properties (Wood and Stankovic, 2005):

- **Malicious:** it is carried out to prevent the network from fulfilling its intended functions. It is not accidental. Otherwise it is not in the domain of security but reliability and fault tolerance.

- **Disruptive:** it degrades the quality of services offered by the network.
- **Asymmetric:** the attacker puts in much less effort compared to the scale of the impact made on the network.

Every networking service may be subject to a DoS attack. In this section we will review important DoS scenarios for ad hoc and sensor networks.

DoS in The Physical Layer

All the physical attacks explained in Section 8.2.1 can also be perceived as DoS attacks because they prevent a network from performing its expected functions. In this section, the physical layer indicates the OSI layer responsible for representing 1s and 0s correctly in the wireless medium, and a DoS attack in the physical layer, which is called *jamming*, means a security threat against this.

A malicious device can jam a wireless carrier by transmitting a signal at that frequency. The jamming signal contributes to the noise in the carrier and its strength is enough to reduce the signal-to-noise ratio below the level that the nodes using that channel need to receive data correctly. Jamming can be conducted continuously in a region, which thwarts all the nodes in that region from communication. Alternatively, jamming can be done temporarily with random time intervals, which can still very effectively hamper the transmissions.

DoS in The Link Layer

The algorithms in the link layer, especially MAC schemes, present many exploitation opportunities for DoS attacks. For example, *MAC layer DoS attacks* such as the following may continuously jam a channel:

- Whenever an RTS signal is received, a signal that *collides with the CTS signal* is transmitted. Since the nodes cannot start transmitting data before receiving the CTS, they continue sending RTS signals.
- If the MAC scheme is based on sleeping and active periods, *jamming only the active periods* can continuously block the channel.
- *False RTS or CTS signals with long data transmission parameters* are continuously sent out, which makes the other nodes that do virtual carrier sensing wait forever.
- *Acknowledgement spoofing*, where an adversary sends false link layer acknowledgements for overheard packets addressed to neighboring nodes, can also be an effective link layer DoS attack.

More complex DoS attacks can be designed based on MAC layer addressing schemes. For example, in sensor networks, global addressing schemes are not used. Instead, schemes like data-centric routing, attribute-based naming and address reuse can be used. A malicious node can conduct a *sybil attack in the MAC layer* to make the other nodes in the region think that all the addresses available are used. This prevents the nodes from even being a part of the network.

DoS against Routing Schemes

Ad hoc networks are infrastructureless and have special routing challenges, which bear additional opportunities for new types of DoS attack against the network layer protocols for ad hoc and sensor networks. These attacks generally fall into one of two categories (Hu *et al.*, 2005): *routing disruption attacks* or *resource consumption attacks*. Routing disruption attacks aim to make the routing scheme dysfunction, making it unable to provide the required networking services. The goal of resource consumption attacks is to consume network resources such as bandwidth, memory, computational power and energy. Both are denial-of-service attacks and examples of them are listed below (Karlof and Wagner, 2003):

- **Spoofed, altered or replayed routing information:** routing information exchanged among nodes can be altered by malicious nodes to have a detrimental effect on the routing scheme.
- **Hello flood attack** (Karlof and Wagner, 2003): a malicious node may broadcast routing or other information with high enough transmission power to convince every node in the network that it is their neighbor. When the other nodes send their packets to the malicious node, those packets are not received by any node (Figure 8.3).

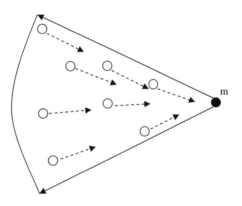

Figure 8.3   Hello flood attack

- **Wormhole attack:** a malicious node can eavesdrop or receive data packets at a point and transfer them to another malicious node, which is at another part of the network, through an out-of-band channel. The second malicious node then replays the packets. This makes all the nodes that can hear the transmissions by the second malicious node believe that the node that sent the packets to the first malicious node is their single-hop neighbor and they are receiving the packets directly from it. For example, the packets sent by node *a* in Figure 8.4 are also received by node *w*1, which is a malicious node. Then node *w*1 forwards these packets to node *w*2 through a channel which is out of band for all the nodes in the network except for the adversaries. Node *w*2 replays the packets and node *f* receives them as if it was receiving them directly from node *a*. The packets that follow the normal route, i.e. *a-b-c-d-e-f*, reach node *f* later than those conveyed through the wormhole and are therefore dropped because they do more hops – wormholes are typically established through faster channels. Wormholes are very difficult to detect and can impact on the performance of many network services such as time synchronization, localization and data fusion.
- **Detour attack:** an attacker can attempt to detour traffic to a suboptimal route or to partition the network. Various techniques can be used for this. For example, Hu *et al.* (2005) define a *gratuitous detour attack*, where a node on a route adds virtual nodes to the route such

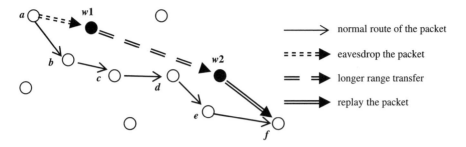

Figure 8.4 Wormhole attack

that the route becomes more costly compared to another route to which the attacker tries to detour traffic.

- **Sink hole attacks:** a malicious node can be made very attractive to the surrounding nodes with respect to the routing algorithm. For example, very attractive routing advertisements can be broadcast and all the neighboring nodes can be convinced that the malicious node is the best next hop for sending the packets to the base station. When a node becomes a sink hole, it becomes the hub for its vicinity and starts receiving all the packets going to the base station. This creates many opportunities for follow-on attacks.
- **Black hole attack:** a malicious node may drop all the packets that it receives for forwarding. This attack is especially effective when the black hole node is also a sink hole. Such an attack combination may stop all the data traffic around the black hole.
- **Selective forwarding (gray hole attack):** when a malicious node drops all the packets, this may be detected easily by its neighbors. Therefore, it may drop only selected packets and forward the others.
- **Routing loop attack:** detour or sink hole types of attack can be used to create routing loops to consume energy and bandwidth as well as disrupting the routing.
- **Sybil attack:** a single node presents multiple identities to the other nodes in the network. This reduces the effectiveness of fault-tolerance schemes and poses a significant threat to geographic routing protocols. Apart from these services it may also affect the performance of other schemes such as misbehavior detection, voting-based algorithms, data aggregation and fusion and distributed storage.
- **Rushing attack** (Hu *et al.*, 2005)**:** an attacker disseminates route request and reply messages quickly throughout the network. This suppresses any later legitimate route request messages, i.e. nodes drop them, because nodes suppress the other copies of a route request that they have already processed.
- **Attacks that exploit node-penalizing schemes:** schemes that avoid low performance nodes can be exploited by adversaries. For example, malicious nodes can report error messages for a node which is actually performing well. Therefore, the routing scheme may avoid using a route that includes this node. Similarly, a link may be jammed for a short time but since error messages are generated about the link during that time interval, the routing scheme may continue to avoid the link even though it is not jammed any more.
- **Attacks to deplete network resources:** when nodes are unattended and rely on their onboard resources, those resources may be depleted by malicious actions. This is especially the case for sensor networks. For example, a malicious node may continuously generate

packets to be sent to the data-collecting node, i.e. the base station, and the nodes that relay these messages deplete their energy.

DoS in The Transport Layer

Transport layer protocols are also susceptible to security threats. Some attack scenarios applicable at this layer are listed below:

- **Transport layer acknowledgement spoofing:** false acknowledgement or acknowledgement with large receiver windows (see Section 6.2) may make the source node generate more segments than the network can handle, causing congestion and degrading the network capacity.
- **Replaying acknowledgement:** in some transport layer protocols, such as TCP-Reno, acknowledging the same segment multiple times indicates negative acknowledgement. A malicious node can replay an acknowledgement multiple times to make the source node believe that the message was not delivered successfully.
- **Jamming acknowledgements:** a malicious node can jam the segments that convey acknowledgements. This may lead to the termination of a connection.
- **Changing sequence number:** in protocols like RMST and PSFQ, a malicious node may change the sequence number of a fragment and make the destination believe that some fragments have been lost.
- **Connection request spoofing:** a malicious node can send many connection requests to a node, using up its resources such that it cannot accept any other connection request.

This list of scenarios is not exhaustive. Many different tactics can be developed based on the protocol used in the transport layer.

DoS in The Application Layer

Application layer protocols can also be exploited in DoS attacks. We mentioned many of them in Section 8.1.2.2. Protocols like node localization, time synchronization, data aggregation, association and fusion can be cheated or hindered, as explained in that section. For example, a malicious node that impersonates a beacon node and gives false location information or cheats with regard to its transmission power, i.e. transmitting with less or more power than it is supposed to do, may hamper the node localization scheme. Since these kinds of attack diminish the related network service, they can also be categorized as DoS attacks.

8.1.2.4 Misbehaving

Note that DoS attacks may sometimes come from nodes within the network. Some nodes may *misbehave* to gain unfair shares of the limited networking resources, i.e. they may employ *selfishness*. For example, by using the MAC scheme, a misbehaving node can force the other nodes into longer back offs and free the network resources for its own use. Nodes may also be selfish by refusing to relay others' messages. If every node acts like this, then selfishness may have an impact similar to a DoS attack.

Another means of misbehaving may be aimed at a charging scheme by denying payment for services received. At first glance ad hoc networks can be thought of as free-of-charge environments where everybody collaborates to communicate with each other through a license-free channel. This is not always exactly the case. Mesh networks provide wireless multihop access to broadband services. Similarly, there can be multihop cellular networks where nodes are allowed to access the network through ad hoc multihop wireless links when they are out of the coverage area provided by the infrastructure, as shown in Figure 8.5. In both of these cases nodes reach a service provider and are supposed to pay for the services they get from the provider. In Salem *et al.* (2003), several attacks are envisaged against the charging schemes in these kinds of network:

- **Refusal to pay:** the source node may deny that it carried out communications specified on a bill.
- **Dishonest rewards:** in multihop networking, intermediate nodes should relay the packets of others. To motivate intermediate nodes to forward the packets of others instead of being selfish, rewarding mechanisms, such as paying them, can be designed. In this case, a misbehaving node may want to appear that it was involved in forwarding some packets, even though it was not.
- **Free riding:** intermediate nodes on the route between the source and destination can piggyback their packets on to ongoing communications to avoid paying the bill. For example, routing node *A* can piggyback its packet to routing node *B* onto packets from the source to the destination in Figure 8.5.

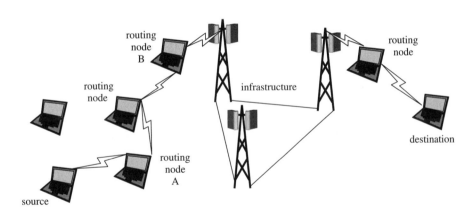

Figure 8.5 Multihop cellular networks

8.2 Attackers

Attackers can also be categorized according to many criteria. Our classification of attackers is based on the characteristics shown in Figure 8.6: emission, location, quantity, motivation, rationality and mobility. First, an attacker can be passive or active; this matches the

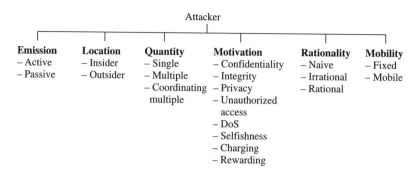

Figure 8.6   Classification of attackers

classification of attacks. Active attacks are carried out by active attackers and passive attacks by passive attackers.

An attacker can be an insider or an outsider. An insider is a node that has been compromised or tampered with, and it is a part of the attacked network. The attacker knows all the cryptographic information owned by the compromised node when it is an insider. Therefore, stealthy active attacks can be organized by insider attackers. Outsider attacks can be either passive or active. In other words, an insider can be perceived as a legal entity inside the network such as a node that has been registered or a node that is allowed to access the network. An outsider is typically a node that is not welcome on the network.

There may be a single attacker or more than one. When there are multiple attackers, they can collaborate with each other, which can be considered a more difficult case to defend against. In Hu *et al.* (2005) active attackers are denoted Active-n-m, where n is the number of insider nodes, and m is the total number of insider and outsider nodes. They then propose an attacker hierarchy with increasing strength as follows:

Active-0-1: the attacker owns only one outsider node.
Active-0-x: the attacker owns x outsider nodes.
Active-1-x: the attacker owns x nodes and only one of them is an insider.
Active-y-x: the attacker owns x nodes and y of them are insiders.

Note that in this hierarchy all the nodes represent a single attacker. Therefore, they are supposed to collaborate.

An adversary carries out attacks with a certain motivation, such as breaking confidentiality, integrity and privacy. This may also be done to gain access to unauthorized resources. An attacker may also attack to hinder the operations of the other side. Selfishness, avoiding payment or getting unearned rewards may be other motives. These have already been explained in the sections on attacks.

Needlessness, malfunctioning nodes and naïve users may also become threats to a network. However, needlessness is not the only reason for 'irrational' attacks – those where the results of the attack may not be worth the cost of attacking. An attacker may attack simply in order to attack and break a security system, perceiving this as a challenge to prove himself/herself.

Rational attackers carry out their attacks to obtain something which is worth more than the cost of the attack.

Finally, attackers can be fixed or mobile. Detecting mobile attackers and defending against them is generally more difficult than defending against a fixed adversary.

8.3 Security Goals

In short, the goal of security is to provide security services to defend against all the kinds of threat explained in this chapter. Security services include the following:

- **Authentication:** ensures that the other end of a connection or the originator of a packet is the node that is claimed.
- **Access control:** prevents unauthorized access to a resource.
- **Confidentiality:** protects overall content or a field in a message. Confidentiality can also be required to prevent an adversary from undertaking traffic analysis.
- **Privacy:** prevents adversaries from obtaining information that may have private content. The private information may be obtained through the analysis of traffic patterns, i.e. frequency, source node, routes, etc.
- **Integrity:** ensures that a packet is not modified during transmission.
- **Authorization:** authorizes another node to update information (import authorization) or to receive information (export authorization). Typically, other services such as authentication and integrity are used for authorization.
- **Anonymity:** hides the source of a packet or frame. It is a service that can help with data confidentiality and privacy.
- **Nonrepudiation:** proves the source of a packet. In authentication the source proves its identity. Nonrepudiation prevents the source from denying that it sent a packet.
- **Freshness:** ensures that a malicious node does not resend previously captured packets.
- **Availability:** mainly targets DoS attacks and is the ability to sustain the networking functionalities without any interruption due to security threats.
- **Resilience to attacks:** required to sustain the network functionalities when a portion of nodes is compromised or destroyed.

8.4 Review Questions

8.1   What are the differences between sybil and masquerading attacks?

8.2   What are the differences between sink hole, wormhole, hello flood and black hole attacks?

8.3   Classify physical active attacks and discuss which can be more threatening to sensor networks.

8.4   Are ad hoc networks or sensor networks more susceptible to traffic analysis? Why?

8.5   What happens if a malicious node alters the sequence numbers of fragments in PSFQ? Does the result change in RMST if it is attacked with the same approach? Why?

8.6   Explain a misbehavior attack in the MAC layer.

8.7   Is SMAC or CSMA/CD more susceptible to misbehavior attacks? Why?

8.8   What are the differences between jamming and EMP attacks?

8.9   Which one has a bigger impact: attacking the time synchronization or node localization scheme? Why?

8.10 Do you think that the security scheme for routes between the source and the infrastructure (from a sensor to the base) differs from the one between the infrastructure and the destination (from the base to a sensor)? Why?

8.11 Give two examples of irrational and two examples of rational attackers. Do you think that there should be different considerations when designing a security system for rational and irrational attackers? Why?

9

Cryptography

Cryptographic systems are typically categorized into two classes – *symmetric* and *asymmetric* – based on the number of keys used in the system. In symmetric cryptography there is a single key known to both sender and receiver, and the same key is used both for encrypting and decrypting a message. In asymmetric systems, separate keys are applied for encryption and decryption.

Cryptographic systems can also be classified as *block* or *stream* ciphers according to the way that they treat the input. In block ciphers one block of elements is processed at a time. For example, the first 64 characters in plain text are processed together, then the second 64 characters and so on. In stream ciphers, each element, e.g. each character, is processed separately as it arrives.

In this chapter, we discuss the fundamentals of cryptography, i.e. symmetric, asymmetric, block and stream as well as hash functions and the use of hash chains and hash trees in providing authentication services.

9.1 Symmetric Encryption

Symmetric (private/secret/single) key cryptography uses one key which is shared by both sender and receiver (Figure 9.1). It is the oldest available technique and was the only one available before the publication of public key cryptography in 1976.

Substitution and *transposition* (permutation) are the two primitives used in symmetric encryption. Substitution ciphers can be grouped into two classes: *monoalphabetic* and *polyalphabetic*.

A monoalphabetic substitution cipher maps a plain text alphabet to a cipher text alphabet, so that each letter of the plain text alphabet maps to a single unique letter of the cipher text alphabet. Caesar's cipher is a good example of a monoalphabetic substitution cipher. In Caesar's cipher every letter in the plain text is replaced by a letter a fixed number of positions shifted in the alphabet. For example, when an alphabet is shifted 4 to the left to obtain the cipher alphabet, 'A' in plain text is replaced by 'E', as shown in Example 9.1.

Security in Wireless Ad Hoc and Sensor Networks   Erdal Çayırcı and Chunming Rong
© 2009 John Wiley & Sons, Ltd

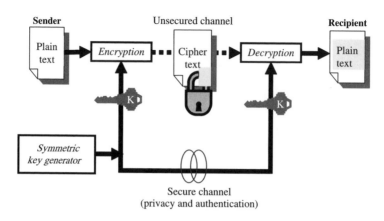

Figure 9.1 Symmetric cryptography

Example 9.1

Caesar's cipher by shifting 4 to left

Plain Alphabet: A B C D E F G H I J K L M N O P Q R S T U V W X Y Z

Rotated four positions to left

Cipher Alphabet: E F G H I J K L M N O P Q R S T U V W X Y Z A B C D

Plain text: HELLO WORLD

Cipher text: LIPPS ASVPH

A monoalphabetic substitution cipher can be broken easily and cannot provide any security.

A polyalphabetic substitution cipher uses a separate monoalphabetic substitution cipher for each successive letter of plain text, depending on a key. The Vigenere cipher is a polyalphabetic cipher where n cipher alphabets are created by shifting the plain alphabet. Each cipher alphabet is denoted by a letter in the plain alphabet based on the number of shifts to obtain the cipher alphabet. For example the 'A' alphabet is the nonshifted version of the plain alphabet and the 'B' alphabet is the one-shifted version of the plain alphabet. For encryption, a key is provided. The first letter in the plain text is replaced by the letter in the cipher alphabet indicated by the first letter in the key. For example, if the key is 'SECURE', the first letter in the plain text is replaced by using 'S' in the cipher alphabet and the second letter in the plain text is replaced by using 'E' in the cipher alphabet, as depicted in Example 9.2.

Example 9.2

Vigenere cipher

Plain Alphabet: A B C D E F G H I J K L M N O P Q R S T U V W X Y Z

A Cipher Alphabet: A B C D E F G H I J K L M N O P Q R S T U V W X Y Z
B Cipher Alphabet: B C D E F G H I J K L M N O P Q R S T U V W X Y Z A
C Cipher Alphabet: C D E F G H I J K L M N O P Q R S T U V W X Y Z A B

Z Cipher Alphabet: Z A B C D E F G H I J K L M N O P Q R S T U V W X Y

Key: SECURE

Plain text: HELLO
Cipher text: ZINFF

Transposition ciphers are also known as *permutation* ciphers. The technique hides a message by rearranging the letter order. In other words, the actual letters used remain in the cipher text. Columnar transposition, explained in Example 9.3, is an example of a transposition cipher.

Example 9.3

Columnar transposition cipher

Plain text: BURN THE LETTER AFTER READING
Key: LUCKY

Key	L(3)	U(4)	C(1)	K(2)	Y(5)
	B	U	R	N	T
	H	E	L	E	T
	T	E	R	A	F
	T	E	R	R	E
	A	D	I	N	G

Read the plain text into cipher text in the column order. The letters in the key also give the order of reading the columns. Since C in the key, i.e, 'LUCKY' is the letter that has the highest precedence within the key, start reading from the C column.

Cipher text: RLRRINEARNBHTTAUEEEDTTFEG

The main weakness of these primitive techniques is their association with natural language characteristics (for example, letter frequency). A cipher needs to obscure the statistical properties of the original message. *Diffusion* and *confusion* are the two methods used to obscure statistical properties. In diffusion, the statistical structure of plain text is dissipated over the bulk of the cipher text. This is achieved by having each plain text element affect the value of many cipher text elements, which is equivalent to saying that each cipher text element is affected by many plain text elements. In a cipher that has a good diffusion property, when a single bit is changed in the plain text, the probability that each cipher text bit changes is 0.5. In confusion, the relationship between the cipher text and the key is made as complex as possible in order to thwart attempts to discover the key. Typically, substitution provides confusion and transposition provides diffusion.

In a *product cipher*, two or more basic ciphers are performed in sequence in such a way that the final result or product is cryptographically stronger than any of the component ciphers. Rotor machines from the Second World War used a product cipher which implemented a very complex and varying substitution cipher. A harder cipher can be made by using several ciphers in succession: two substitutions make a more complex substitution, two transpositions make a more complex transposition, but a substitution followed by a transposition makes a new, much harder cipher.

Shannon introduced the idea of *substitution–permutation networks* based on two primitives: substitution and permutation. They provide confusion and diffusion of messages. Modern block ciphers are based on Shannon's substitution–permutation network concept of an invertible product cipher.

In symmetric encryption, it is essential that the sender and receiver have a secure channel to exchange secret keys. There is also a need for a strong encryption algorithm. What this means is that if someone has a cipher text, a corresponding plain text message and the encryption algorithm, they still cannot determine the key or decrypt another cipher text. In other words, someone who has a plain text for a given cipher text and the algorithm should not be able to break the cipher.

Brute force and *cryptanalysis* are two methods of breaking an encryption. Brute force checks all possible combinations and eventually determines the plain text message. If the key space is very large, this becomes impractical. Cryptanalysis is a form of attack that attacks the characteristics of the algorithm to deduce a specific plain text or the key used. One would then be able to figure out the plain text for all past and future messages that continue to use this compromised set-up. A cipher text-only-attack relies only on analysis of the cipher text itself by applying various statistical tests to it. In a known plain text, the analyst may be able to deduce the key based on knowledge of the way in which the known plain text is transformed; the knowledge comes from analysis of one or more plain text messages captured along with their encryptions. In a chosen-plain text-attack, the analyst is able to choose the messages to encrypt; hence, deliberately picked patterns can be selected in order to reveal the structure of the key.

Although a designer would like to make his/her algorithm as difficult as possible to cryptanalyze, there is great benefit in making the algorithm easy to analyze. That is, if an algorithm can be concisely and clearly explained, it is easier to analyze the algorithm for cryptanalytic vulnerabilities and therefore develop a higher level of assurance as to its strength.

Many widely used encryption algorithms, such as Data Encryption Standard (DES), Triple DES (3DES), Advanced Encryption Standard (AES) and the Rivest Cipher (RC5), are symmetric block product ciphers. Most of these symmetric block ciphers have a similar structure to the Feistel Cipher where many rounds of basic ciphers are performed in sequence in such a way that the final result or product is cryptographically stronger than any of the component ciphers. The Feistel Cipher is proven to have a good *avalanche effect*, which means that a small change in either the plain text or the key produces a significant change in the cipher text.

In the Feistel Cipher the input is divided into blocks of $2n$ bits and each block is encrypted by using a key K, as depicted in Figure 9.2. First the right half, i.e. the least significant n bits, of the $2n$ bit input block has a function F applied to it, where K provides the parameters for the function, and then the exclusive OR (XOR) of the result and the most significant n bits of the input block is taken for substitution. After this, a transposition is performed by interchanging

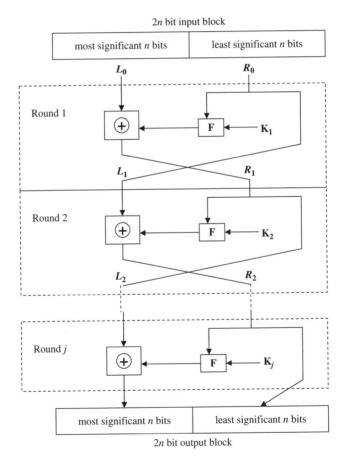

Figure 9.2   Feistel Cipher

the least significant n bits of the input block and the result of XOR. This is one round of the Feistel Cipher, where the same operation is applied over j rounds to the output of the previous rounds, as shown in Figure 9.2. In each round a new key K_i is also derived from the key K_{i-1} of the previous round.

DES has been the most widely used encryption algorithm since it was adopted as the federal information processing standard in the US in 1977. It is almost the same as the original Feistel network shown in Figure 9.2. There are 16 rounds of processing, and in each round a subkey specific for the round and generated from a 56-bit-long original key is used. For decrypting a cipher text, the same process is applied but the subkeys are used in reverse order. DES was considered an efficient encryption scheme and implemented as hardware and software for many applications. However, its key length is the main weakness. Crackers have broken DES by the brute force technique.

3DES provides more resilience to brute force attacks. It executes DES three times and uses three keys, i.e. a different key for each of these three executions. It encrypts plain text by using the first key K_1, decrypts the result of the first execution by using the second key K_2 (i.e. it applies the subkeys from K_2 in the reverse order during the second execution) and finally

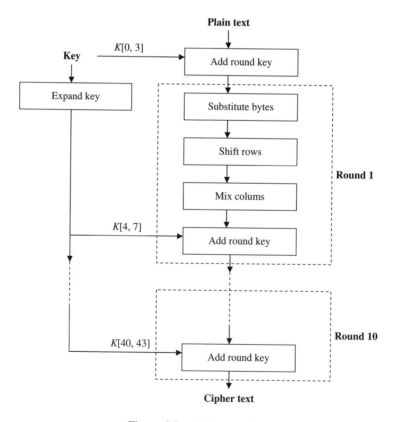

Figure 9.3 AES encryption

encrypts one more time the result of the second execution by using the third key K_3. The result of this third execution is the cipher text transferred to the receiver. For decryption this sequence is applied in the reverse order. First the cipher text is decrypted by using K_3, the result of this stage is encrypted by using K_2 and finally the result of the second execution is decrypted by using K_1. This is a low-cost and compatible fix for the weaknesses of DES. It provides a 168-bit key length and uses the same data encryption algorithm (DEA) as in DES. This DEA has been subject to more cryptanalytic attack attempts than any other algorithm. Therefore, the level of confidence for DEA is high. Moreover, the most widely used encryption hardware and software can also be used for 3DES.

However, DEA, and therefore 3DES, also have some other drawbacks. First is the block size, which is 64 bits. Larger block sizes are better for both efficiency and security. Second, DEA was designed in the 1970s, and there are algorithms that may better fit the software implementation based on contemporary generic hardware. Therefore, a new standard called AES has been adopted since 2001. AES has major differences from the Feistel structure. In Figure 9.3, the AES encryption steps are shown. Decryption follows exactly the same steps and uses the same keys in the reverse order.

Symmetric cryptography is much faster to implement than asymmetric cryptography. However, if the secret key is disclosed, communications are compromised. The symmetric method also means that parties are equal; hence it does not protect a sender from a receiver forging a message and claiming that the message was sent by the sender.

9.2 Asymmetric Encryption

The problem with secret keys is how to exchange them securely over public networks like the Internet. The communication is only secure if the secret key involved remains secret, but to establish a secure channel to convey keys over the public medium is costly and often impractical. Asymmetric cryptography, also known as public key cryptography, was introduced in 1976 (Diffie and Hellman, 1976) and offers a solution to this challenge. The idea is to make the encryption and decryption keys different, so that knowledge of one key would not allow a person to find out the other. Requirements for the realization of public key cryptography are given by Diffie and Hellman (1976):

- a computationally easy algorithm can generate a pair of keys, i.e. a private key K_S and a public key K_P;
- a plain message M can be easily encrypted to a cipher message M_C by using the public key K_P;
- the cipher message M_C can be easily decrypted to the plain message M by using the private key K_S;
- the cipher message M_C cannot be decrypted to the plain message M by using the public key K_P;
- an adversary that knows the public key K_P cannot determine the private key K_S.

There are two widely used algorithms for public key cryptography: RSA and Diffie–Hellman. We explain RSA (Rivest *et al.*, 1978), developed by Ron Rivest, Adi Shamir and Len Adleman, in Example 9.4.

Example 9.4

RSA Algorithm

Public and private key selection	**Example**
1. Select two prime numbers p and q, $(p \neq q)$	$p = 7$, $q = 13$
2. Calculate $n = p \times q$	$n = 7 \times 13 = 91$
3. Calculate $\Omega = (p - 1) \times (q - 1)$	$\Omega = 6 \times 12 = 72$
4. Select a prime number e, $e < \Omega$	$e = 5$
5. Calculate d, $d \times e \bmod \Omega = 1$	$d = 29$
6. Public key $K_P = \{e, n\}$	$K_P = \{5, 91\}$
7. Private key $K_S = \{d, n\}$	$K_S = \{29, 91\}$

Encrypting plain text M	**Example**
M must be smaller than n, $M < n$	$M = 2$
$M_C = M^e \bmod n$	$M_C = 2^5 \bmod 91 = 32$

Decrypting cipher text M_C	**Example**
$M = M_C^d \bmod n$	$M = 32^{29} \bmod 91 = 2$

Public key cryptography is asymmetric since the parties are not equal. A user's private key is kept private and known only to the user. The user's public key is made freely available to others to use. Each encryption/decryption process requires at least one public key and one private key. As shown in Figure 9.4, the public key can be used to encrypt information that can only be decrypted by the possessor of the private key.

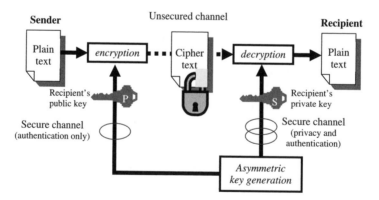

Figure 9.4   Asymmetric cryptography

The private key can also be used to encrypt a signature, as shown in Figure 9.5. For example, a digest of a message can be encrypted by using the private key and appended to the message as a digital signature. The receiver that has the public key can decrypt the digital signature. The receiver also generates the digest by using the same hash function and can compare that

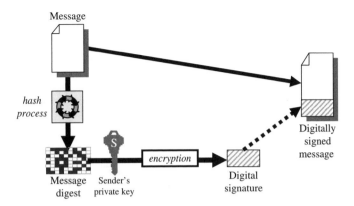

Figure 9.5 Digital signatures

to the received digest. If they are identical, that proves the identity of the sender because only the sender that has the private key can encrypt the digest such that it can be decrypted by using the public key. Digital signatures are useful both for authentication (Figure 9.6) and nonrepudiation services.

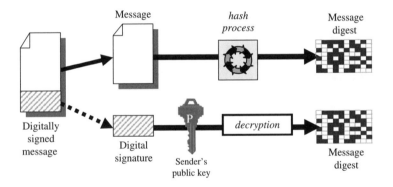

Figure 9.6 Authentication using digital signatures

Authentication and privacy can be achieved in combination, as shown in Figure 9.7. After a digital signature is added to a message by using the sender's private key, it can also be encrypted by using the receiver's public key. Hence, only the intended receiver can decrypt the message by using the receiver's private key. This provides privacy. After decrypting the message, the receiver also generates the digest and compares it with the digital signature by decrypting it with the sender's public key, and this latter process provides authentication and nonrepudiation.

Public keys must be authentic even though they do not need to be secret. Otherwise a user could pretend to be another user, i.e. masquerade, and broadcast a public key. *Digital certificates* (or simply certificates) are used for providing authentic public keys in asymmetric cryptography (Figure 9.8). A digital certificate, which contains a public key and additional

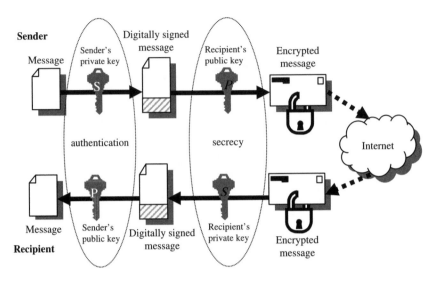

Figure 9.7   Authentication and secrecy using public key cryptography

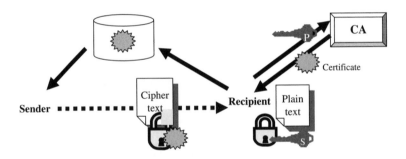

Figure 9.8   Classical use of asymmetric cryptography

information about the key owner, including its identification, is created and signed by a *certificate authority* (CA). A participant, i.e a receiver, sends his public key to the CA and is given a certificate with the matching private key. The receiver publishes the certificate via a directory, i.e. a server, or sends its certificate to the sender. The other participants, i.e. senders, can verify that the CA created the certificate by using the CA's signature in the certificate and the CA's public key. A sender uses the public key in the certificate to encrypt a message, which can be decrypted only by using the associated private key in the receiver.

X.509 is the international standard for providing public key certificate authentication services. The certificates are placed in a directory so that users may easily obtain the certificates of other users. As the number of users increases, a single directory may become insufficient. Moreover, there may be multiple CAs. Therefore, CAs are arranged in a hierarchy in X.509, which also provides the standards for the format of certificates. X.509 also introduces a revocation service in case certificates are compromised. This is managed with a certificate revocation list (CRL) signed by the CA that issues the certificate.

Certificate management is an overhead that may not always be affordable and practical for every system. Therefore, *identity-based encryption* (IBE) (Boneh and Franklin, 2001) has been introduced, where the public key can be derived from an arbitrary string such as the user's ID, the role name, the name of a group of users, etc. (Figure 9.9). A private key generator (PKG) issues the corresponding private key. Thus, IBE eliminates the need for large databases to maintain the correspondence between an identity and the related public key, which is necessary in certificate-based PKC solutions. This simplifies the key management, saves space and ensures that attacks on the certificate database are no longer a threat. It is also referred to as *certificateless public key cryptography*.

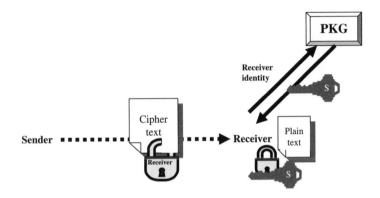

Figure 9.9   Use of identity-based cryptography

In IBE public keys are arbitrary strings, or 'identities', such as names, roles, email addresses, etc. A trusted private key generator (PKG) using a master secret derives private keys from the identities, i.e. IBE public keys. The master secret is necessary to compute private keys. Therefore, users cannot collude to obtain the private keys of other users. Nodes obtain their private keys from the PKG. If a node needs to send a message to another node, it encrypts the message by using the receiver's identity. The receiver can decrypt this message by using only the private key generated for its identity by the PKG. Therefore, a secure communication is achieved. One main concern in IBC is that one must put total trust in the PKG to generate private keys.

Public-key cryptography (PKC) introduces a means of secure communication over a public channel. One problem with asymmetric encryption, however, is that it is slower to implement than symmetric encryption. It requires far more processing power for encryption and decryption compared to symmetric cryptography. Therefore, asymmetric key cryptography complements rather than replaces private key cryptography. The most common use of PKC is in combination with symmetric encryption. PKC is often used to agree on a secret key, and the secret key is then used for symmetric encryption.

9.3 Hash Functions and Message Authentication Code

A cryptographic hash function is based on running a one-way compression function from a block of data of any size to an output of fixed length n. This happens in many rounds, and the output size n implies the level of security provided by the hash function. A hash function should be easily computable and publicly known, making both hardware and software implementations practical. It should also have the following additional properties:

- **One way property:** when a hash value $H = h(M)$ for a message M is given, it is computationally infeasible to find out the message M.
- **Weak collision resistance property:** for any given block M, it is computationally infeasible to find $M' \neq M$ with $h(M') = h(M)$.
- **Strong collision resistance property:** it is computationally infeasible to find any pair (M, M') such that $h(M') = h(M)$.

Hash functions view an input as a sequence of n bit blocks and process them iteratively one block at a time. A simple hash function is depicted in Example 9.5. Note that this simple hash function always generates an n-bit code independent of the size of input.

Example 9.5

A simple hash function using bitwise XOR

$$C_i = C_{1i} \oplus C_{2i} \oplus C_{3i} \oplus \ldots \oplus C_{mi}$$

Input stream: 11100101011100010101010101100010

Code length n = 6

```
Block 1:   1  1  1  0  0  1
Block 2:   0  1  0  1  1  1
Block 3:   0  0  0  1  0  1
Block 4:   0  1  0  1  0  1
Block 5:   1  0  0  0  1  0
Code:      0  1  1  1  0  0
```

Hash functions can use a block cipher or modular operations as their compression function. However, the use of a block cipher is less efficient due to the key–subkey processing. The focus of most people's attention are dedicated hash functions like the MD (message digest) family, represented by MD2, MD4 and MD5; the secure hash algorithm (SHA) family, represented by SHA-0, SHA-1, SHA-256, SHA-384 and SHA-512; the RIPEMD (RACE integrity primitives evaluation message digest) family, represented by RIPEMD, RIPEMD-128 and RIPEMD-160; and others like HAVAL and Whirlpool. MD5 makes only one pass over the data and generates a 128-bit digest. It has been proved many times that MD5 cannot satisfy the collision resistance property. In 2006, an algorithm that could find collisions for MD5 digests within minutes by using a simple portable computer was published (Klima, 2006).

SHA-1, depicted in Example 9.6, is the secure hash algorithm that replaced SHA-0. It is widely used in many cryptographic systems. As is shown in Example 9.6, it processes the data in 512-bit blocks and in four rounds. Each round has 20 steps where a function, i.e. F, and a constant value, i.e. K, specific to the round are applied to the values in the message digest buffers. SHA-1 always produces a message digest that has a length of 160 bits.

Example 9.6

SHA-1

```
Notation:  ∨ (bitwise or operation)
           ∧ (bitwise and operation)
           ! (bitwise negation)
           ⊕ (bitwise xor operation)
           ← (bitwise left rotate operation)
           | append
```

Preprocess the input

Append the bit '1' to the input message

Append k bits '0' to the input message, where $k \geq 0$ and the resulting message length in bits is 64 bits less than a multiple of 512 bits.

Append the length of message in bits before preprocessing as a 64-bit big endian integer.

Initialize message digest buffer

```
H0=0x67452301
H1=0xEFCDAB89
H2=0x98BADCFE
H3=0x10325476
H4=0xC3D2E1F0
```

Process message in 512-bit blocks until the end of message

Assign message digest buffer values

```
A=H0
B=H1
C=H2
D=H3
E=H4
```

Break the block into 16 words of 32 bits, i.e.

$w[i], 0 \leq i \leq 15$

Extend the block to 80 words, i.e.

```
For i from 16 to 79
```
$$w[i] = (w[i-3] \oplus w[i-8] \oplus w[i-14] \oplus w[i-16]) \leftarrow 1$$

Round 1

K=0X5A827999

```
For i from 0 to 19
    F=(B∧C)∨(!B∧D)(there are alternative functions)
    T=(a←5)+f+E+K+w[i]
    E=D
    D=C
    C=B←30
    B=A
    A=T
```

Round 2

K=0X6ED9EBA1

```
For i from 20 to 39
    F=B ⊕ C ⊕ D
    T=(a←5)+f+E+K+w[i]
    E=D
    D=C
    C=B←30
    B=A
    A=T
```

Round 3

K=0X8F1BBCDC

```
For i from 40 to 59
    F=(B∧C)∨(B∧D)∨(C∧D)(there are alternative functions)
    T=(a←5)+f+E+K+w[i]
    E=D
    D=C
    C=B←30
    B=A
    A=T
```

Round 4

K=0XCA62C1D6

```
For i from 60 to 79
    F=B ⊕ C ⊕ D
    T=(a←5)+f+E+K+w[i]
    E=D
```

```
        D=C
        C=B←30
        B=A
        A=T
```

Add this block's hash result to the message digest buffer
```
        H0=H0+A
        H1=H1+B
        H2=H2+C
        H3=H3+D
        H4=H4+E
```

When all message ends, output the message digest (MD)
```
        MD=H0|H1|H2|H3|H4
```

A hash function like SHA-1 produces a single fixed-length output from two fixed-length inputs. The length of the output is typically the same as the length of one of the inputs. In SHA-1 the total length of the values used for initializing the message digest buffers is 160 bits, and the output is also 160 bits independent of the length of the other input, which is at least 512 bits. Therefore, these hash functions are also called compression functions that transform one large input into a shorter, fixed-length output.

Hash functions provide a measure of data integrity. For a message M, hash code $H = h(M)$ is computed and sent with the message. If the hash code $H' = h(M')$ computed by the receiver by using the same hash function h for a received message M' is different from H sent with the message, then the receiver must accept that M' is a modified version of M. Here, H is called the *message integrity code* (MIC) and it should be transmitted encrypted to be used as a reliable gauge of the message integrity.

On the other hand, a *message authentication code* (MAC) that uses a secret key as an input in the compression process does not need to be encrypted. An example of a MAC algorithm is *hash message authentication code* (HMAC). HMAC is designed such that it can use any other available hash function, such as MD5 or SHA-1. The design objectives of HMAC are as follows:

- to use any available hash function;
- to replace the used hash function easily;
- to introduce negligible overhead in addition to the overhead by the used hash function;
- to present a clear cryptographic analysis of the authentication strength.

In the HMAC structure, as shown in Figure 9.10, the key is first padded with zeros on the left such that it becomes b bits long. The padded key K^+ is then applied via a XOR operation to *ipad*, which is 00110110 repeated $b/8$ times. The result is appended to the message. This makes the input for a selected hash function. In the second round, K^+ is applied via a XOR operation to another constant *opad*, which is 01011100 repeated $b/8$ times, and appended to the result of the first round. This final bit stream has the hash function applied one more time, which produces an m-bit long MAC.

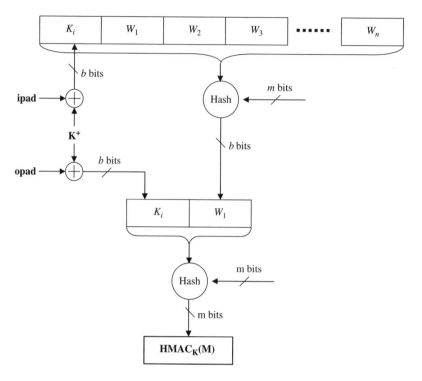

Figure 9.10   HMAC

In summary, there are three ways to produce MACs:

1. Produce an MIC by using a cryptographic hash function and append it to the message after encrypting by using a symmetric encryption algorithm. This provides both integrity and authentication.
2. Produce a MAC by using a structure like HMAC and append it to the message transferred. This also provides both integrity and authentication.
3. Produce an MIC by using a cryptographic hash function and append it to the message after encrypting by using an asymmetric encryption algorithm, i.e. digital signatures. This provides nonrepudiation in addition to integrity and authentication.

9.4 Cascading Hashing

Cascading hashing involves the composition of several hash functions into a single one. The goal is usually to increase security. Given hash functions h and g, a natural construction is simply to form $H = (h(M)|g(M))$, which turns two n-bit hash functions into a $2n$-bit one. However, this simple construction does not provide much additional security. There are more secure methods of cascading hashing, such as hash chains and trees.

9.4.1 Hash Chains

A hash chain is generated by repeatedly applying a hash function h to a string M. In Figure 9.11, a hash chain of length three is shown.

$$X_2 = h(M)$$
$$X_1 = h(h(M)) = h^2(M) \qquad (9.1)$$
$$X_0 = h(h(h(M))) = h^3(M)$$

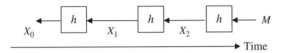

Figure 9.11 A hash chain of length three

The sender may use this hash chain in reverse order for authentication. A receiver initially stores X_0. At a later time, the sender may disclose X_1 and the receiver can verify X_1 by checking $h(X_1) = X_0$. Similarly, the following packages can be verified by releasing the later hash values in the chain. This approach is used by TESLA for the authentication of broadcast or multicast messages. We elaborate on TESLA later in this chapter.

9.4.2 Hash Trees

Hash trees, suggested by Ralph Merkle, are also known as Merkle trees. A hash tree is a tree of hashes in which the leaves are hashes of data blocks, e.g. data blocks in a file or a set of files. Nodes further up in the tree are the hashes of their respective children.

For example, in Figure 9.12, H_0 is the result of hashing $H_{0,0}$ and then $H_{0,1}$:

$$H_0 = h(H_{0,0}|H_{0,1}) \qquad (9.2)$$

A hash tree is often implemented as a binary tree; in other words, each node has two child nodes. However, more children nodes for each node can also be allowed. Usually, a cryptographic hash function such as SHA-1 or Whirlpool is used for hashing. At the top of a hash tree there is a *top hash* (or *root hash* or *master hash*).

One of the main applications of hash trees is authentication of a file downloaded in a P2P network. Before downloading on a P2P network, the top hash of the file is acquired from a trusted source. Then, the hash tree can be obtained from any non-trusted source, like any peer in the P2P network, and the received hash tree is checked against the trusted top hash. If the hash tree is damaged or fake, another hash tree from another source will be tried until the program finds one that matches the top hash.

One branch of the hash tree can be downloaded at a time and the integrity of each branch can be checked immediately, even though the whole tree is not available yet. This can be an advantage since it is more efficient to split files into smaller data blocks so that only small

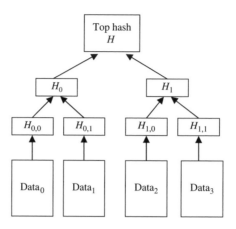

Figure 9.12   A hash tree of height 3

blocks have to be retransmitted if they get corrupted during transmission. For the case of a large file, its hash tree is also relatively large. However, one small branch can be downloaded quickly, the integrity of the branch can be checked and then the other data blocks can be downloaded.

9.4.3 TESLA

A broadcast authentication protocol enables receivers to verify that a received packet was really sent by the claimed sender. In other words, such a protocol should prevent impersonation of the sender so that any receiver can easily verify that the data were sent by the legitimate sender. Simple association of each packet with a message authentication code (MAC), where a checksum is computed using a shared secret key, can be used for authentication in a two-party communication. However, it does not provide secure broadcast authentication. This is because every receiver in the broadcast network has the secret key and can use it to forge data and impersonate the sender. Asymmetric cryptography with digital signatures has promising functions for meeting the requirements of broadcast authentication, but it is not a good solution in practice because of the associated overhead and relatively longer time and higher bandwidth requirements.

An alternative approach to providing authentication in broadcasting is still to use MACs and symmetric cryptography, but based on delayed disclosure of keys by the sender. In Perrig *et al.* (2000a) and Perrig and Tygar (2003), *TESLA* (timed efficient stream loss-tolerant authentication) is proposed as one such scheme. It is further given as an IETF draft (Perrig *et al.*, 2000b; 2003) of a multicast source authentication technique.

In the TESLA approach, each message packet broadcast by the sender is appended with a MAC generated using a secret key k, known only originally to the sender; the sender will disclose the key k to the receiver after a certain time delay d. The receiver buffers the received packet without being able to authenticate it until receiving the key. After the specified time delay d, the sender discloses k and the receiver is then able to authenticate the buffered data packet using the disclosed key k. A single MAC per packet is sufficient to provide authentication. One requirement of this approach is synchronization between the sender and the receiver.

In TESLA, a one-way key chain is used to provide authentication. A one-way key chain is generated by repeatedly using the same one-way hash function on an initial key. For instance, in Figure 9.13, the sender generates the chain by randomly selecting k_n and repeatedly applying the one-way function h. Then the sender commits k_0 to a receiver. Any element of the chain can later be verified by the sender through the chain starting with k_0 to reveal the values in the opposite order.

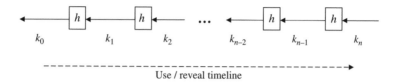

Use / reveal timeline

Figure 9.13 Generation and use of one-way hash key chain

TESLA includes four phases: sender set-up, receiver set-up, message broadcast and message authentication at the receiver.

In the sender set-up phase, the sender first divides the time into intervals and assigns one key from the one-way key chain to every time interval. The one-way key chain is computed by

$$K_n = F(K_{n+1}) \tag{9.3}$$

where F is a one-way function. The one-way key chain is committed to the network and is used in the reverse order of generation. The sender defines a disclosure time interval d after which the value will be published.

In the receiver set-up phase, the receiver should get loosely time synchronized with the sender. Then the receiver obtains the key disclosure schedule from the sender using an authenticated channel. The key disclosure schedule includes the interval duration, start time, index of interval, the length of the one-way key chain, the key disclosure delay d and a key commitment to the key chain K_i.

After the sender and receiver set-up phases, the sender broadcasts a message. The procedure for this is depicted in Figure 9.14. The key K_i' used to generate the MAC is generated from key K_i using a one-way function F'. Every time the sender broadcasts the message, it adds the MAC generated using the key corresponding to the time interval when the message is broadcast. The sender broadcasts the related one-way chain value after d. A message should include the following fields:

$$P_j = \{M_j || \mathrm{MAC}(K_i', M_j) || K_{i-d}\} \tag{9.4}$$

When a receiver receives a packet, it first checks that the key used to compute the MAC is still secret by comparing the time it received the packet and the time for the sender to disclose the secret key. If the MAC key is still secret, then the receiver buffers the packet. After the

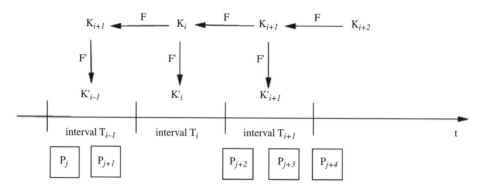

Figure 9.14   Sending authenticated packets

receiver receives the secret key for this packet, the disclosed key is first verified and then used to verify the buffered packet that was sent in the time interval of the disclosed key. If the MAC is correct, the receiver accepts the packet.

TESLA was originally designed for broadcast authentication that requires loose time synchronization between the sender and receiver. Since it requires a digital signature to authenticate the initial packet and it needs to store the key chain in its memory, it is impractical to use in sensor networks where nodes only have limited resources. Therefore, μTESLA, explained in Chapter 13, was introduced.

9.5 Review Questions

9.1   What are the primitive techniques used in symmetric encryption?

9.2   What are the main methods used to attack symmetric encryption?

9.3   What are the disadvantages of symmetric encryption?

9.4   The following cipher text is the result of Caesar's cipher. Find the plain text.

KNSIRJ

9.5   A spy uses a product cipher whose encryption consists first of a row transposition cipher and second a Caesar cipher, and the keys are located in the address of a house, where the house telephone number is the key to the row transposition cipher and the house number is the key to the Caesar cipher. According to the schedule, today's keys are associated with John Smith in the phone book:

John Smith, Lagårdsveien 3, 4010 Stavanger Tel: 51 63 47 82.
Today's message to the spy is:
HSQQD XHDRP YFKWV HNHDL OULLQ DDWVW BDWWA RJULS
What is the clear-text of this message?

9.6   How can authentication and privacy be achieved together using asymmetric encryption?

9.7   How does identity-based encryption work and what are the advantages of identity-based encryption?

9.8   Generate a public key and a private key for $p = 17$ and $q = 11$ by using RSA. Encrypt 52 and decrypt the result by using the keys that you generate.

9.9   What are the characteristics required for a hash function?
9.10   What are the differences between MAC and digital signatures? Discuss the advantages
       and disadvantages of them with respect to each other.
9.11   What is a hash chain and how can it be used?
9.12   What is a hash tree and how can it be used?
9.13   How does TESLA work and what are its advantages?
9.14   Why is TESLA not appropriate for wireless sensor networks? Discuss the techniques
       that can be used to adapt TESLA for wireless sensor networks.

10

Challenges and Solutions: Basic Issues

There are a number of security challenges associated with wireless ad hoc and sensor networks, and we introduced them in Chapter 7. In contrast to guided media, the wireless medium is openly accessible, less reliable and has no obvious physical boundary. An attacker does not need to break any physical barriers to gain access to the wireless medium and can enter the network from anywhere and from all directions. In addition, more complications in security establishment come from the dynamically changing topology, the reliance on node collaboration for network connectivity, the lack-of-trust infrastructure and the lack of a clear line of defense. Since ad hoc networks are built without a fixed infrastructure and centralized management, the protection mechanisms used in tethered networks cannot be adopted directly in wireless ad hoc and sensor networks.

In a secure wireless ad hoc sensor network, a node is authorized by the network and only authorized nodes are allowed to access the network resources. The generic process to establish such a network consists of bootstrapping, pre-authentication, network security association establishment, authentication, behavior monitoring and security association revocation. Among these, authentication is of the utmost importance and is an essential service in network security. Other basic security services like confidentiality, integrity and nonrepudiation depend on authentication. Secret information is exchanged only after nodes are able to verify and validate one another.

Bootstrapping (often shortened to booting) is the phase in which the nodes in a network are made aware of the presence of all or some of the others in the network. During the bootstrapping process, all nodes that want to join the network must gain their identifying credential to prove their eligibility to access the protected network. The identifying credential takes the form of either something that they should have or something that they should know. For instance, such a credential may be a global network key or an updated list of trusted nodes. Upon bootstrapping completion, the network should be ready to accept the participation of any node with a valid credential. A joining node must present its credential to the network in order to prove its eligibility to access protected resources or offered services.

Security in Wireless Ad Hoc and Sensor Networks   Erdal Çayırcı and Chunming Rong
© 2009 John Wiley & Sons, Ltd

All nodes need to share identifying credentials to be able to prove their identity to each other. This initial exchange of credentials is called *pre-authentication*. Once credentials are verified, network security associations are established with other nodes and these security associations become further proof of authorization in the network. The nodes in the network can be associated securely by, for instance, a symmetric key, a public key pair, a commitment of a hash key chain or some contextual information. The security association will expire after a prenegotiated time and can be renegotiated after expiry. Communications between nodes can now be authenticated by using the security associations.

Using a security association, a node can be authenticated and then the behavior of the node can be monitored by the network in order to discover compromised or misbehaving nodes. If the network finds that a node has been compromised by an adversary, the node is isolated from the network by getting its network associations revoked or its reestablishment of association request denied. In most scenarios, network security association establishment and revocation are implemented under network key distribution/exchange and management.

10.1 Bootstrapping Security in Ad Hoc Networks

In the bootstrapping phase the nodes in a network become aware of the presence of the other nodes in their vicinity or in the overall network. Wireless ad hoc networks introduce new challenges for this phase. An important characteristic of wireless ad hoc networks is the lack of centralized security infrastructure. To protect the security of a network, the first step is to build a security infrastructure between the nodes during the bootstrapping phase. The trust infrastructure should satisfy the requirements that only legitimate nodes can join the network; new nodes that may join the network can form a secure association with the nodes already in the network; the trust infrastructure can be set up without the knowledge of the network topology; the credential verification scheme should be strong enough to resist DoS attacks and at the same time should not require large computational ability and memory.

To build such a trust infrastructure, prior context can be used. If nodes have a shared prior context before the deployment of the network, they can use this information to enter the network. For example, a secret key can be predistributed if the nodes were initiated from the same trusted environment. However, this assumption is not always practical.

A trusted third party could also be used to facilitate the establishment of such an infrastructure. The trusted third party may be a certification authority (CA), a base station or a designated node. All nodes in the network have to agree on a trusted third party. In addition, trust services are centralized in the network. Hence, if there is a central and stable node like a base station in the network, such a node can be converted into a trusted third party node. However, this is not well suited to most ad hoc settings, where nodes are distributed and no natural centralized trusted candidates can be found.

In practice, in wireless ad hoc networks, the topology of the network changes quickly and therefore it is difficult to get either a trusted prior context or a trusted third party. For ad hoc networks, it is more natural to self-organize the trust infrastructure since this involves no special nodes, no infrastructure, no centralized configuration point and no shared prior context. However, an out-of-band authenticated communication channel is often needed in many proposed protocols. For example, in Balfanz *et al.* (2002) a privileged side channel is used to exchange public information to help the node perform the pre-authentication protocol for bootstrapping secure communication in an ad hoc wireless network. Identity-based security schemes are another approach to achieving this goal. Alternatively, a key can be established with a tamper-proof hardware token provided by users.

10.2 Bootstrapping Security in Sensor Networks

The characteristics of wireless sensor networks, such as limited battery power and lack of infrastructure, make them more vulnerable to attack than conventional ad hoc networks. In addition to the bootstrapping challenges explained for conventional ad hoc networks, the following are introduced for wireless sensor networks (Chan *et al.*, 2003):

- **Resilience to node capture:** sensor nodes are easy targets for capture in many deployment scenarios. If a node is captured, a physical attack on a sensor node may reveal the secret information in the memory. Resilience to node capture should be part of the security system. In addition to this, the security system should be so resilient that communication between captured nodes and noncaptured nodes cannot be compromised.
- **Resistance to node replication:** by node capture or infiltration, an adversary may obtain secret information enabling replication of a legitimate node, and then gain full control of the network by populating it with clones to such an extent that the legitimate nodes are outnumbered.
- **Revocation:** a misbehaving node should be dynamically removed from the system once it has been detected.
- **Scalability:** as the number of nodes in a network grows, the security characteristics mentioned above may be weakened. Security mechanisms for sensor networks should be capable of accommodating large numbers of sensors and allowing new sensors to be added to the network.
- **Memory and energy performance efficiency:** security mechanisms should have few long-term and dynamic memory requirements, low computational demands and even lower communication demands.

For bootstrapping, a sensor node can start with the minimum output power level and send a 'Hello' message to discover neighbors in the vicinity. Then it can increase the output power level to discover the other nodes, which are not within range when the minimum transmission power is used. This can be repeated with an increased transmission power each time until a given number of neighbors are discovered or the maximum transmission power level is reached. A similar strategy, called *incremental shouting*, is applied in Subramanian and Katz (2000). LEACH (Heinzelman *et al.*, 2000), explained in Chapter 5, can also be perceived as a method of bootstrapping, where nodes can declare themselves as cluster heads based on a certain probability, and nodes access the closest node declared a cluster head. These techniques and others proposed as MAC or routing protocols can be used to discover neighboring nodes. When they are not secured, they provide opportunities for adversaries to introduce insider attacks. To secure the bootstrapping process, key distribution, exchange and management play an important role.

10.3 Key Distribution, Exchange and Management

In an ad hoc network, the trustworthiness of a communicating node is crucial. For secure data exchange, a secure association is often established by setting up shared credentials, e.g. a secret key, between neighboring nodes. To establish the security association, key management protocols, including key distribution and key exchange protocols, have core importance in bootstrapping. After bootstrapping, an ad hoc network is initialized and ready to accept any participant with a valid credential. In other words, the possession of a valid credential becomes proof of the trustworthiness of a newly joined node.

The text from the next paragraph until the end of Section 10.3 is © 2006 IEEE and is reprinted with permission from Hegland, A M, Winjum, E, Mjølsnes, S F, Rong, C, Kure, Ø and Spilling, P. 'A Survey of Key Management in Ad Hoc Networks,' in the *IEEE Communications Surveys & Tutorials*, ISSN 1553-877X, pp. 48–66, Vol. **8**(3), 3rd quarter issue of 2006.

There are also other related surveys of key management schemes. A survey of key distribution mechanisms for wireless sensor networks is in Camtepe and Yener (2005). Key management schemes for secure group communication are surveyed in Rafaeli and Hutchison (2003). Reviews of key management protocols for ad hoc networks and sensor networks can be found also in Fokine (2002); Djenouri *et al.* (2005); Law (2005) and Merwe *et al.* (2005).

The desirable features of an ad hoc network key management scheme:

- **Applicability:** the various key management schemes focus on different targets. The aim may range from group key establishment to availability of central management entities. Their applicability depends on the fundamental assumptions as to network origin (planned or truly ad hoc), network size, node mobility, geographic range and the required level of human involvement.
- **Security:** authentication and intrusion tolerance is a primary concern to ensure no unauthorized node receives key material that can later be used to prove status as a legitimate member of the network. Nobody should provide private keys or issue certificates for others unless the others have been authenticated. Intrusion tolerance means system security should not succumb to a single, or a few, compromised nodes. Other central security issues are trust management and vulnerability. Trust relations may change during network lifetime. The system should enable exclusion of compromised nodes. In order to judge the security of a key management scheme, possible vulnerabilities should be pinpointed. Proper key lengths and cryptographic algorithms of adequate strength are assumed.
- **Robustness:** the key management system should survive despite denial-of-service attacks and unavailable nodes. The key management operations should be able to be completed despite faulty nodes and nodes exhibiting Byzantine behavior; that is, nodes that deliberately deviate from the protocol. Necessary key management operations caused by dynamic group changes should execute in a timely manner. Key management operations should not require networkwide and strict synchronization.
- **Scalability:** key management operations should finish in a timely manner despite a varying number of nodes and node densities. The fraction of the available bandwidth occupied by network management traffic should be kept as low as possible. Any increase in management traffic reduces available bandwidth for payload data accordingly. Hence, scalability of key management protocols is crucial.

- **Simplicity:** simplicity regarding user-friendliness and communication overhead is an additional intuitive and overall critical factor to the success of a key management scheme. We reckon, however, that a system that is secure, robust and scalable implies simplicity. Given that these conditions are fulfilled, we believe simplicity is first and foremost a matter of implementation.

The ideal key management service for ad hoc networks should be simple, formed on the fly, never expose or distribute key material to unauthorized nodes, ensure that system security does not succumb to (a few) compromised nodes, easily allow rekeying / key updates, enable withdrawal of keys when nodes are compromised or keys for other reasons should be revokable, be robust to Byzantine behavior and faulty nodes, scale well enough to handle the expected network sizes and node densities and efficiently manage network splits and joins.

Signed routing information requires a security association that allows one-to-many signing and verification. Routing messages are often broadcast, and all receiving nodes should be able to check the validity. Messages such as neighbor-detection messages are not forwarded by other nodes. Other routing messages, such as topology information messages in proactive routing protocols and route requests and route replies in reactive routing protocols, are flooded into the entire network. The receiving nodes may not be known to the transmitting node. In addition, bandwidth is limited. Unique signatures for each receiver scale badly. In other words, pairwise keys provide no good option for protection of routing information.

10.3.1 Standards

None of the emerging MANET (mobile ad hoc network) Internet drafts and RFCs has thus far encompassed key management. Of other standards, the IEEE 802.11i (ANSI/IEEE, 2004) security amendment for IEEE 802.11 wireless local area networks assumes keys are preshared or established with the aid of fixed infrastructure. In the case of truly ad hoc communication, preshared symmetric keys are the only option. The aim of IEEE 802.11i is protection of payload (data frames) on layer 2. IEEE has, in 2005, begun work on 802.11w that will cover security on management frames. Other standards for wireless communication include the ZigBee (ZigBee Alliance, 2004)/IEEE 802.15.4 (IEEE-SA Standards Board, 2003) and the Bluetooth (Bluetooth SIG, 2004) specifications for personal area networks. The preconditions of these standards are infrastructure-based networks and do not apply to MANETs. ZigBee specifies key management for the security elements of IEEE 802.15.4. ZigBee assumes the initial keys are predistributed, installed out of band or received in the clear over the air from a trust center. Keys in Bluetooth are derived with the aid of PIN codes. A common PIN code is entered out of band in pairs of nodes that wish to communicate.

10.3.2 Classification of Key Management Schemes

We can classify key management into two families: *contributory*, where all nodes take part equally in the key management; and *distributive*, where a cluster head conducts the key management alone.

In contributory key management schemes, every node will contribute to the generation and distribution of the cryptographic keys. In other words, key management is based on the contributions of all the nodes in the network as a collaborative effort. Some of the contributory schemes studied here rely on a centralized entity, others do not. This approach is particularly appropriate for a network with a small number of nodes and can offer strong security properties such as key independence and forward secrecy.

In distributive schemes, each key originates from a single node. The nodes may still cooperate during the key distribution. Distributive schemes may be centralized but can also be distributed. In the latter, each node generates a key and tries to distribute it to others. Distributive schemes may involve one or more trusted entities and comprise both public key systems and symmetric systems. Public key schemes include traditional certificate-based and identity-based schemes. The symmetric schemes are classified as either MANET schemes or WSN (wireless sensor network) schemes.

The classification categories 'contributory' and 'distributive' reflect best the origin of the keys in the schemes. The classification is illustrated in Figure 10.1.

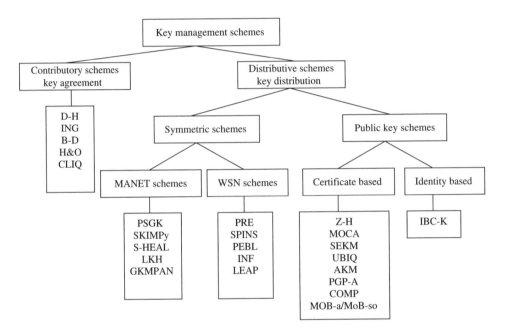

Figure 10.1   Classification of key management schemes

10.3.3 Contributory Schemes

Contributory schemes are characterized by the lack of a trusted third party responsible for generation and distribution of the cryptographic keys. Instead, all communicating parties cooperate to establish, i.e. 'agree' upon, a secret symmetric key. The number of participants ranges from two parties (establishing a pairwise key) to many parties (establishing a group key). Although not necessarily designed with ad hoc networks in mind, intuitively the contributory approach of collaboration and self-organization may seem to fit the nature of ad hoc networks. A number of contributory schemes are reviewed and evaluated in this section. Only one of these was designed specifically for ad hoc networks.

10.3.3.1 Diffie–Hellman (D–H)

D–H (Diffie and Hellman, 1976) establishes a unique symmetric key between two parties. It relies on the discrete log problem (DL); deciding S given $g^\wedge S \bmod p$ being a hard problem. D–H is outlined in Figure 10.2. The parties agree upon a large prime, p, and a generator, g.

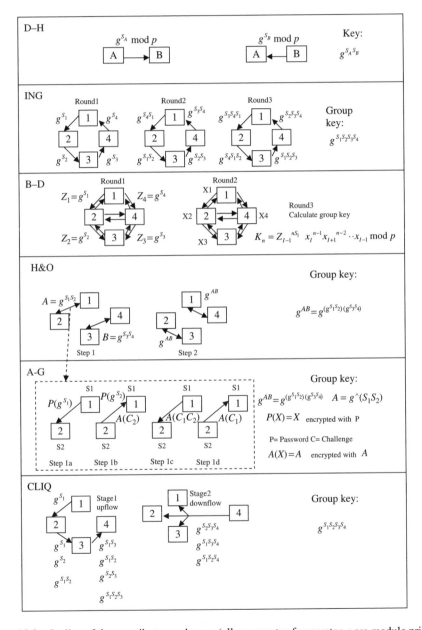

Figure 10.2 Outline of the contributory schemes (all exponents of generator g are modulo prime p)

Each party randomly chooses a secret S_A and S_B and transmits the public values, $(g^\wedge S_A)$ *mod p* and $(g^\wedge S_B)$ *mod p*, as shown in the figure. Raising the number received from the other party to the power of its own secret, gives a common secret key, $g^\wedge(S_A S_B)$ *mod p*, shared only by the two.

Like any scheme involving pairwise unique keys, D–H provides intrusion tolerance. A captured node only compromises the keys it shares with its communicating peers. Byzantine and faulty nodes basically only disturb their own key establishment with communicating peers. D–H is vulnerable to man-in-the middle (MIM) attacks. It is left for the nodes to judge who to trust. But as authentication is lacking, Alice cannot be sure that she actually communicated with Bob and not Charlie.

The generic D–H scheme is not applicable for protection of routing information in ad hoc networks. It applies to two parties only. Protection of routing messages with pairwise keys necessitates a different signature for each possible recipient, which scales badly. Based on the D–H scheme, the majority of the contributory schemes seek to remedy the shortcomings of D–H regarding MIM vulnerability and extendibility to more than two parties.

10.3.3.2 Ingemarsson, Tang and Wong (ING)

ING (Ingemarsson *et al.*, 1982) provides a symmetric group key by extending the two-participant D–H scheme to n participants. Figure 10.2 shows the principles with four nodes. All nodes are arranged into a logical ring. After $n - 1$ rounds, each node can calculate the secret key. Each round involves an exponentiation from every node, and every node must transmit its share to the next node in the logical ring, as shown in the figure.

ING lacks authentication and is vulnerable to MIM attacks. It scales poorly. Communicational complexity grows proportionally to the number of nodes squared. Byzantine behavior or faulty nodes may inhibit successful key establishment. A captured node means the group key is compromised and necessitates a rekeying. The scheme does not specify how compromised nodes can be detected. The requirement for the nodes to organize into a logical ring during the key agreement procedure makes ING unsuitable for ad hoc networks. The establishment of keys for protection of routing information implies a logical ring of one-hop neighbors only (all nodes within direct transmission range). With mobile nodes and unstable links it is questionable whether ING will ever complete successfully.

10.3.3.3 Burmester and Desmedt (B–D)

B–D (Burmester and Desmedt, 1994) seeks to establish a group key. It relies on the DL problem. But, contrary to the other contributory schemes studied here, it is not based on D–H. An outline of B–D with four nodes is shown in Figure 10.2. B–D completes in three rounds. Every node picks a secret, S_i, and multicasts its *public value*, $Z_i = g^\wedge S_i$ to all other nodes in the group. In round 2, every node calculates and multicasts a new public value. This value is derived by dividing the public value received from the next node by the public value received from the previous node in the logical ring of nodes, and raising the result to the power of its own secret S_i, as illustrated in Figure 10.2. In the third and final round every node calculates the conference key from its secret and the information received from all the other nodes in the previous rounds.

B–D is apparently more efficient than ING as it completes in three rounds. However, each round requires a high number of exponentiations and reliable multicasting. Reliable multicasting is difficult in tethered networks and even more challenging in ad hoc networks. Changes in group membership necessitate a restart of the key-agreement procedure. In an ad hoc network with moving nodes it may thus never be possible to establish a group key by B–D, nor handle later changes in group membership. Group changes will certainly cause delay and disruption. B–D also demands an already running routing protocol or only one-hop neighbors, i.e. the key agreement schemes depend on an already established routing infrastructure – but the infrastructure cannot be established before the keys have been set up. B–D authentication of the public values (not shown in the figure) can be implemented with the aid of predistributed public keys. Trust is managed through the certificate issuer. This implies a planned network and the basic key management problem reverts to a public key scheme.

10.3.3.4 Hypercube and Octopus (H&O)

H&O (Becker and Wille, 1998) reduces the number of rounds and exponentiations of ING from n to d ($n = 2^d$) by arranging the nodes in a *hypercube*, i.e. a d-dimensional cube. Figure 10.2 illustrates H&O in a network with four (2^2) nodes. In step 1, nodes 1 and 2 perform a D–H key agreement. Nodes 3 and 4 do the same. The symmetric keys established in step 1 are used as the secret values in a new D–H key agreement in step 2: nodes 1 and 4 perform a D–H key agreement and nodes 2 and 3 do the same and so on. H&O actually consists of two protocols: Hypercube and Octopus. Hypercube assumes the number of participants is a power of 2. Octopus extends Hypercube to allow an arbitrary number of nodes.

H&O is vulnerable to MIM attacks as authentication is absent. Byzantine or faulty nodes may preclude successful key agreement. Changes in group membership require rekeying. It is left for the nodes to decide when rekeying is needed. Like B–D and ING, H&O relies on an underlying communication system to provide a consistent node-ordering view to all group members. Besides the difficulty of keeping a consistent node ordering where nodes join and leave dynamically, it implies an already running (unprotected) routing protocol or only one-hop neighbors. The latter scales badly. Altogether, H&O is unsuitable for network layer security in ad hoc networks.

10.3.3.5 Password Authenticated Key Agreement (A-G)

A-G (Asokan and Ginzboorg, 2000) is the only one of the contributory systems studied that has been designed with ad hoc networks in mind. A-G is basically H&O extended with password authentication, as indicated in Figure 10.2. It assumes all legitimate participants receive a password offline (written on the conference hall blackboard or distributed through another location limited channel). The nodes must prove knowledge of the password during the pairwise D–H key agreements of the H&O protocols, as shown in Figure 10.2. The figure shows the password-authenticated key agreement between two nodes. The password is used to encrypt the public value and an initial challenge in a challenge–response protocol, as illustrated in the figure.

A-G doubles the number of messages and increases the computational complexity compared to H&O. It remedies H&O's vulnerability to MIM attacks at the price of scalability.

A-G inherits the deficiencies of H&O regarding dependability on an already established communication infrastructure and node-ordering scheme. Hence, it is not suitable for network layer security in mobile ad hoc networks.

10.3.3.6 CLIQUES (CLIQ)

CLIQ (Steiner *et al.*, 1998; 2000) is outlined in Figure 10.2. It extends the generic D–H protocol to support dynamic group operations. CLIQ distinguishes between initial key agreement (IKA) and auxiliary key agreement (AKA). IKA takes place at group formation. AKA handles all subsequent key agreement operations. In both cases, a group controller synchronizing the key agreement procedure is required.

The figure shows the IKA protocol with four nodes. Stage 1 (the *upflow* stage) starts from node 1 which picks a secret exponent, S_1, and unicasts $g^\wedge S_1$ to the next node. Node 2 picks a secret exponent S_2, and unicasts to node 3 the values shown in the figure. The procedure is repeated until the final node – the *group controller* – is reached. The group controller is now able to calculate the secret group key, i.e. the generator g raised to the power of the secret exponents of all nodes in the group. In stage 2 (the *downflow* stage) the group controller multicasts the intermediate values required by each of the other nodes to calculate the secret group key, as shown in the figure.

Both AKA (not shown in the figure) and IKA rely on the group controller. The group controller of CLIQ thus represents a single point of failure. Each AKA operation results in a new group key that is independent of all previous keys. Adding a new member with AKA basically extends stage 1 of the IKA protocol with one node. The role of the group controller can be fixed or floating. Allowing any node to take over the role as group controller renders the system vulnerable to malicious nodes. CLIQ omits authentication. The designers have left security properties such as authentication out while focusing on group changes, but argue that authentication could easily be added. Other major drawbacks with CLIQ, as with B–D, are dependency upon reliable multicast and availability of a consistent view of node ordering. With variable connectivity it is questionable whether IKA and AKA would ever complete successfully. With unstable links, highly mobile nodes and rapid splits and joins, instability may result.

10.3.3.7 Other Contributory Schemes

A large number of key agreement schemes relying on already distributed keys have been proposed. The basic key management problem thus reverts to distribution of the initial keys. Several schemes are also two-party protocols unsuitable for network layer security and are therefore left out of further discussions. Examples include MQV (Certicom Corp., 2004) based on traditional public keys, schemes relying on identity-based public keys, such as Chen and Kudla (2003) and Wang (2005) and the D–H based protocols proposed in Cagalj *et al.* (2006).

10.3.3.8 Summary of the Contributory Key Management Schemes

The main implications and limitations of various types of contributory schemes in ad hoc networks are demonstrated by the schemes studied in this section. Although the contributory approach at first glance may seem to fit the self-organizing nature of ad hoc networks, none of the contributory schemes are good candidates for key management in ad hoc networks. D–H, ING and H&O can be skipped due to missing authentication. They are vulnerable to MIM attacks. B–D and CLIQ can be left out – no matter whether the authentication scheme is included or not – as they have an inherent survivability problem with dependency on reliable multicasting. A-G fails on scalability and robustness due to a dependency upon node ordering and availability of all nodes during group changes.

10.3.4 Distributive Schemes

Distributive schemes involve one or more trusted entities and comprise both public key systems and symmetric systems. Truly ad hoc networks require the trusted entity to be established impromptu during network initialization. The distributive category is divided into symmetric and public key schemes.

10.3.4.1 Public Key Schemes

Certificate-based public key schemes require the public keys to be distributed in a way that allows the receiving nodes to verify the authenticity of the key material. The wired network solution is a public key infrastructure (PKI) where a centralized certificate authority (CA) issues certificates binding the public keys to specific users/nodes.

If it is suspected that a node has fallen into the wrong hands, or the node for other reasons should be expelled, the certificate is revoked. Revoked certificates are added to the certificate revocation list (CRL). The CA signature guarantees the authenticity of certificates and CRLs. Under the assumption that a centralized trusted entity is not well suited for ad hoc networks, where overall availability cannot be guaranteed all the time, the proposed key management schemes for ad hoc networks involving certificate-based PKI advocate various ways to distribute the CA functionality. The intuitive approach of naive CA replication is not reckoned good enough as it poses poor intrusion tolerance. With more nodes holding the private CA key, there is a higher risk of getting it compromised.

Partially Distributed Threshold CA Scheme (Z-H)

Z-H (Zhou and Haas, 1999) assumes a PKI system and puts forward a framework to provide an available, intrusion-tolerant and robust CA functionality for ad hoc networks. The private CA key is distributed over a set of *server* nodes through a (k,n) *secret sharing scheme* (Shamir, 1979). The private CA key is shared between n nodes in such a way that at least k nodes must co-operate in order to reveal the key. (Finding the private CA key S is comparable to finding $f(0)$ given a polynomial $f(x)$ of degree $k-1$ and knowing k values, e.g. $f(1), f(2)\ldots f(k)$.)

When queried, each server generates a partial signature of the certificate using its private key share in a *threshold signature scheme* (Desmedt, 1994). A server acting as combiner collects the partial signatures and produces a valid signed certificate.

Z-H advises *share refreshing* to counter *mobile adversaries*, i.e. adversaries that temporarily compromise one server and then attack the next. *Proactive secret sharing schemes* (Herzberg *et al.*, 1995) allow the shareholders to periodically refresh their shares through collaboration. An adversary must thus compromise more than t shares *between* refreshes in order to compromise the system. The original secret does not change, only the shares held by the servers. (Bear in mind the homomorphic property: If $(s_1, s_2 \ldots, s_n)$ is a (k,n) sharing of S and $(a_1, a_2 \ldots, a_n)$ is a (k,n) sharing of A, then $(s_1 + a_1, s_2 + a_2, \ldots, s_n + a_n)$ is a secret sharing of $S + A$ (Zhu *et al.*, 2005). Choosing $A = 0$ gives a new sharing of S. The scheme is made robust to missing and erroneous shares through *verifiable secret sharing* (Pedersen, 1991): extra public information testifies to the correctness of each share without disclosing the share.

Although not clearly stated, the system relies on a central trusted *dealer* to bootstrap the key management service and decide which nodes shall act as servers. Z-H assumes an underlying (unsecured) routing protocol.

According to Zhou and Haas (1999) nodes cannot get the current public keys of other nodes or establish secure communication with others if the CA service is unavailable. However, every node should hold a copy of its own certificate. For network layer security, it would be more efficient to receive the certificates directly from the communicating peers (or other nodes in the neighborhood). If the certificate is needed to verify a signature on routing information, the node in question must certainly be available; otherwise there would be no requirement to verify its routing message. Thus, the need for online CA access is limited. Every node must contact the CA to get its initial certificates (and receive the public key of the CA). The same is true if the node for some reason has lost its private key or has had its certificate revoked. However, to get a new certificate, the node should be authenticated by the CA service – which necessitates some sort of physical contact between the node and the CA service. Certificate *updates* call for CA service. For scenarios like emergency and rescue operations, it would be better to make sure certificates are renewed in the preparation phase and not during network operation.

The CA service is needed for revocation and distribution of CRLs. Z-H postulates that public keys of nodes that are no longer trusted, or have left the network, should be revoked. In an ad hoc network it can be hard to decide when a node has actually left the network. Revoking keys due to temporal missing connectivity would not be wise. More important is revocation of keys belonging to captured nodes. The frequency of such revocations in networks for emergency and rescue operations will expectedly be low.

Periodical share refreshing implies some form of synchronization. Synchronization is bandwidth consuming and difficult in ad hoc networks. Management traffic between server nodes and certificate exchanges also consumes much bandwidth, and makes Z-H scale badly. A single CA or hierarchy of CAs is likely to prove better than the Z-H approach. Efficient spreading of the cross-certificates in the respective domains is a problem for further investigation. There is no easy way to update the private/public CA key pair and make sure all nodes are informed.

MOCA

MOCA (Yi and Kravets, 2002a; 2002b) is basically an extension of Z-H (Zhou and Haas, 1999). The focus is on distributed CA services and communication between the nodes and the server nodes – MObile Certificate Authorities (MOCAs). Whereas Z-H does not state how to select CA servers, MOCA suggests the nodes that exhibit the best physical security and computational resources should serve as MOCAs. The MOCA scheme furthermore 'moves' the combiner function of Z-H from the CA servers to the requesting end nodes. The benefit is a less vulnerable scheme as the nodes no longer depend on the availability of the CA server nodes to combine the partial certificate signatures.

A MOCA certification protocol, *MP*, is proposed to provide efficient and effective communication between clients and MOCAs. According to the MP, certificate requests should be unicast to β specific MOCAs that, based on fresh routing entries or short distances, are likely to be accessible. With the (k,n) threshold scheme, k MOCAs are required to complete a certification service. To increase probability of receiving at least k responses: $\beta = k + \alpha$. When availability drops, the protocol returns to flooding (as in Z-H). It is assumed that the MP maintains its own routing tables and co-exists with a 'standard' ad hoc routing protocol.

Secure and Efficient Key Management (SEKM)

In essence, SEKM (Wu *et al.*, 2005) suggests the servers of MOCA form a multicast group. The aim is efficient updating of secret shares and certificates. A node broadcasts a certificate request to the CA server group. The server that first receives the request generates a partial signature and forwards the request to an additional $k + \alpha$ servers (not a true multicast). Only k partial signatures are required. The additional ones are for redundancy in case some are lost or corrupted. SEKM does not state how a server can tell it is the first to receive the refresh request and start the $k + \alpha$ forwarding. On the whole, SEKM has the same features as MOCA. The required number of servers still has to be contacted and the partial signatures returned.

Ubiquitous Security Support (UBIQ)

UBIQ (Kong *et al.*, 2001) is a fully distributed threshold CA scheme. Similar to the partially distributed CA schemes Z-H, MOCA and SEKM, it relies on a threshold signature system with a (k,n) secret sharing of the private CA key. Differently from the partially distributed CA schemes, *all* nodes get a share of the private CA key. A coalition of k one-hop neighbors forms the local CA functionality. It does not require any underlying routing protocol – only a node density of k or more one-hop neighbors. Mobility may help finding the required number of CA nodes. UBIQ prescribes share refreshing.

The nodes earn trust in the entire network when they receive a valid certificate. Any node holding a certificate can obtain a share of the private CA key. A new secret share is calculated by adding partial shares received from a coalition of k neighbors. The first nodes receive their certificates from a dealer before joining the network. After k nodes have been initialized, the dealer is removed. The authors suggest that as the certification service is delivered within one-hop neighborhoods, some reliable out-of-bound physical proofs, such as human perception, can be used to authenticate new nodes.

Limiting CA service requests to one-hop neighborhoods is bandwidth efficient and good for the scalability. A local coalition can decide to let in nodes from different domains. A drawback is the possible requirement of human involvement. In addition, k should be chosen carefully. A low value reduces intrusion tolerance. A large k necessitates many neighbors. Joshi *et al.* (2005) suggest more shares per node to succeed also with less than k neighbors. In effect, this solution gives little else other than reducing the value of k. Distributing the CA functionality boosts the availability of private key shares. Anyone capable of collecting k shares or more can reconstruct the private CA key. Like any public key scheme relying on a trusted entity, there is no easy way to change the private/public CA key pair during operation.

In Capkun *et al.* (2003a) it is argued that UBIQ may succumb to a sybil attack (Douceur, 2002) where a single node takes on more identities. With offline authentication of new nodes and the certificates serving as proof of trustworthiness, this is hardly a realistic threat – at least not in settings like emergency and rescue operations. Secure and efficient revocation is an unresolved challenge.

Autonomous Key Management (AKM)

AKM (Zhu *et al.*, 2005) provides a self-organizing and fully distributed threshold CA. With few nodes in the network, the scheme is parallel to UBIQ. Each node receives a share of the private CA key. As the number of nodes increases, a hierarchy of key shares is introduced. New nodes then receive a share of the private CA key.

The root CA private/public key pair is bootstrapped by a group of neighbor nodes through *distributed verifiable secret sharing* (Gennaro *et al.*, 1999): each of the n neighbors chooses a secret value S_i and distributes secret shares of this to the other neighbors using a (k,n) secret sharing scheme. This approach is contributory in nature. However, derivation of the individual private/public key pairs of the nodes is not. AKM is therefore classified as a distributive scheme. Authentication is added offline. The sum of the individual secret values $S = (S_1 + S_2 + S_3 + .. + S_n)$ represents the private CA key. The corresponding public CA key equals $g^\wedge S$ (operations are mod prime p). Assuming the nodes publish the individual public values, $g^\wedge S_i$, the public key can be derived without revealing the private CA key by multiplying individual values $g^\wedge S = g^\wedge S_1^* g^\wedge S_2^* \ldots^* g^\wedge S_n$. Figure 10.3 shows the principles.

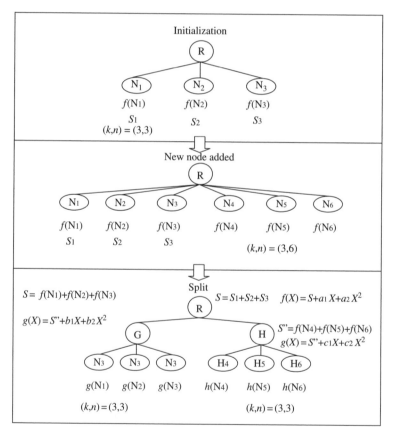

Figure 10.3 The principles of secret sharing in AKM; initialization, node addition and regional splits (all operations are modulo prime p)

The nodes $N_1 - N_6$ and their shares, $f(N_i)$, can be regarded as the leaves of a tree-structure. 'R' in Figure 10.3 is a virtual node representing the private CA key. The probability of a compromise increases with more nodes holding a share of the private CA key. Therefore, when the number of shareholders reaches a certain level, the nodes split into smaller regional groups that set up a new regional key. Before splitting, the nodes $N_1 - N_6$ hold shares $f(N_1) - f(N_6)$ of the private CA key. Assuming the nodes $N_1 - N_3$ decide to form a new group and $N_4 - N_6$ another, N_1 distributes a share of its secret share $f(N_1)$ to the other nodes in the new group. The others do the same with their key shares. The new regional secret of N_1, N_2 and N_3 equals the sum of their shares $S' = f(N_1) + f(N_2) + f(N_3)$ represented as virtual node G in Figure 10.3.

When the number of shareholders in any region reaches the specified level, the region is split. Regions are also merged. With less than k nodes, there are too few nodes to provide a CA service. Certificates signed with regional keys have less assurance than those signed with the CA key. A high-assurance certificate requires partial signatures from nodes in different regions. The scheme assumes the network evolves from the nodes that initiated the AKM service.

In AKM, each node maintains a CRL. AKM does not specify networkwide dissemination of revocation information. A certificate is revoked when at least k neighbors have posted accusations against it. From a security point of view, it is questionable to what extent a certificate signed by the private CA key should be revoked by a group holding only a share of the private CA key.

AKM increases intrusion tolerance at the price of communication cost. Nodes are assumed to disassociate with the previous region and associate with the new when they move from one region of the network to another. Implicitly, the nodes must maintain a view of the key hierarchy and be able to detect regional boundaries. With mobile nodes and unstable links, it is not evident how this can be implemented. The scheme requires the nodes to collaborate on changes in regions and key hierarchy. Byzantine or faulty nodes may delay these operations. In scenarios like emergency and rescue operations, where the CA services primarily are needed for issuance of initial certificates and revocation, a hierarchical AKM with several regions represents a waste of bandwidth. For robustness and scalability, a single region is preferable. The scheme then equals UBIQ.

Self-Organized Key Management (PGP-A)

Capkun, Buttyán and Hubaux (2003a) propose a fully self-organizing key management scheme (PGP-A) – a PGP scheme (Zimmermann, 1994) adapted to ad hoc networks. The CA functionality is completely distributed. All nodes have equal roles. They generate their own private/public key pair and issue certificates to the nodes they trust. Certificates are stored in the nodes rather than in centralized repositories. PGP-A assumes trust is transitive, i.e. if Alice trusts Bob and Bob trusts Charlie, then Alice should also trust Charlie. The nodes merge their certificate repositories and try to find a *verifiable chain* of certificates. The *Maximum Degree* algorithm is suggested to construct a certificate graph with high connectivity even if the sizes of the users' certificate repositories are small – due to the *small world* phenomenon (the hypothesis that everyone in the world can be reached through a short chain of social acquaintances). Certificates are revoked through revocation messages from their issuer, or implicitly revoked at expiry time. Renewals require contact with the issuer. Certificates are also exchanged periodically between neighbor nodes. Evaluation of expiration times and periodical exchanges requires some sort of synchronization between the nodes. It is not evident from the paper how this synchronization should be established.

Periodic certificate exchanges and contact with issuers to have certificates updated is bandwidth consuming and scales badly. PGP-A implicitly requires an already running routing protocol. Trust could be established ad hoc through physical contact and key exchange via a side channel. However, human interaction to keep the network service running is undesirable.

Byzantine behavior or faulty nodes have limited power to prevent others from exchanging certificates. A compromised node only discloses the keys held by this node. Still, a compromised node could be used to issue certificates allowing other illegitimate nodes to gain access to the network.

There is only a probabilistic guarantee that a chain of trust can be found between parties wishing to communicate. On the other hand, trust transitivity combined with the reliance on the small world phenomenon implies that everyone will soon end up trusting everyone. The result is no intrusion resistance. An alternative would be to restrict the maximum number of hops and allow the nodes to differentiate the level of trust they put in the various certificates, as suggested in COMP (Yi and Kravets, 2004).

Composite Key Management for Ad Hoc Networks (COMP)

COMP (Yi and Kravets, 2004) combines MOCA's (Yi and Kravets, 2002a; 2002b) partially distributed threshold CA with PGP-A (Capkun *et al.*, 2003a) certificate chaining. The aim is higher security than obtainable with PGP-A, and increased availability of the CA service compared to MOCA. Nodes that have been certified by the CA are allowed to issue certificates to others. Nodes requesting a certification service should first try the MOCA CAs. If this fails, they should search for neighbors that have been certified by the CA. Depending on configuration, nodes with longer certificate chains to the CA may also be entitled to issue certificates to others.

Each certificate in COMP includes a *confidence value* reflecting the level of confidence the certificate issuer has in the binding between node identity and key (0 = no trust, 1 = full trust). Multiplication of the confidence values gives a measure for the level of trust in a certificate chain. Short certificate chains are generally preferred over long ones. The probability of one or more compromised nodes in the chain grows as the length of the chain increases. Similarly to PGP-A, COMP assumes a level of trust transitivity. However, signing a certificate verifying you believe a key belongs to a certain identity does not necessarily have to mean you also trust this identity to correctly sign certificates of others.

The confidence values enable fine grained evaluation of trust and the nodes do not have to trust the CA fully. However, deciding a proper confidence level is difficult. COMP does not state how the certificate issuers should accomplish this. Byzantine or compromised nodes may, in any case, assign full trust to untrustworthy nodes. Nevertheless, intrusion tolerance is increased compared to pure PGP-A as COMP restricts the maximum length of the certificate chains.

Offline authentication typically includes human interaction, which is cumbersome in the setting of emergency and rescue operations. Interaction with one neighbor is less demanding than the UBIQ requirement for involvement of several neighbors, though. Still, COMP scales no better than MOCA as nodes requesting CA service should first try the MOCA CAs. Transfers of certificate chains limit the scalability additionally.

In MANETs for applications like emergency and rescue operations, the CA will be expected primarily to issue and revoke certificates. Periodical updates of the certificates should not take place online during a rescue operation. Revocation is not addressed by COMP. It is reasonable that the node that issued a certificate is entitled to revoke it. But empowering single ordinary nodes to revoke certificates issued by the CA solely because they hold a certificate signed by the CA renders the system vulnerable to compromised and Byzantine behaving nodes. Allowing a single node to issue certificates contradicts the purpose of the distributed CA.

A search for neighbors certified by the CA in order to obtain an initial certificate requires knowledge of the public CA key. Hence, at some point there should have been an authenticated channel between the searching node and the CA. The initial authenticated channel is typically obtained through physical contact or a short-range side channel. A natural question for the node asked to provide CA service is then: Why did the requesting node not receive its certificate through the authenticated channel simultaneously?

Mobility-Based Key Management Scheme (MOB)

MOB (Capkun *et al.*, 2003b; 2006) seeks to mimic human behavior: if people want to communicate securely, they just get close to each other in order to exchange information. Security associations are established between pairs of nodes that get close. The scheme can be fully self-organizing (*MOB-so*) or rely on an offline authority (*MOB-a*). MOB-so can be based on symmetric or public keys. MOB-a is intrinsically public key based.

A major difference between MOB-so and MOB-a lies in the level of human involvement. In MOB-so, the users should authenticate the communicating peer physically before they establish a security association. The security credentials, *triplets*, are then exchanged over a secure (short-range) side channel. The triplets include *user identifier, key* and *node address*. The nodes also sign and exchange a statement that proves a security association has been established between the two. MOB-so accepts one level of transitivity in trust: security associations can be established through *friends*, i.e. nodes that have security associations to both nodes in question. MOB-a assumes predistributed certificates, and suggests the exchange of security credentials is restricted to one-hop neighborhoods.

In both MOB-so and MOB-a only the keys held by the specific node are compromised when a node is captured. Byzantine behavior or faulty nodes do not inhibit others from exchanging security credentials. The offline authority assumed by MOB-a implies no revocation. The authors suggest compromised nodes should revoke their own certificates. However, it can be hard to tell whether a compromise has taken place or not. Revocation on suspicion represents vulnerability. It may be a threat to availability. Furthermore, if the node has been captured, it may no longer operate according to protocol. With MOB-so, it is left for the user to decide which of its security associations are no longer valid and which friend nodes have turned into enemies.

The MOB schemes are bandwidth efficient in the sense that security credentials are only exchanged within one-hop neighborhoods. Still, the scalability is limited. The MOB schemes imply a long delay to establish security associations with all communication partners. This is also unsuitable for emergency and rescue operations.

MOB-a brings little achievement over predistributed certificates without restrictions on certificate exchanges. Depending on routing protocol, confining certificate exchanges to one-hop neighborhoods may inhibit efficient network formation. There is no security achievement from such a restriction. The signature of the authority ensures the validity of the certificate no matter from whom the certificate was received. The assumption of MOB that no one should communicate securely with parties they have not been close to contradicts the evolution of PKI.

Identity-Based PublicKey (IBC-K)

Identity-based cryptography, introduced by Shamir (1984), removes the need for certificates. Identity-based public key schemes represent a new type of public key system. They allow user identities, e.g. email or IP addresses, to be used as public keys, and make certificates superfluous. A trusted entity is, however, required in order to generate and distribute the private keys corresponding to the various identities. The trusted entity is also needed for revocation. The trusted entity may sign a list of withdrawn identities. As with traditional public key systems, spreading the trusted entity over more nodes has been suggested.

Identities are typically short – at least compared to certificates with a size of several kilobytes. Assuming information that is by default transferred in the routing messages can be used as the public key, identity-based schemes may scale better than the traditional certificate-based approaches. This makes identity-based protocols interesting for bandwidth limited ad hoc networks.

Shamir constructed an identity-based *signature* (IBS) scheme. To verify a signature, it is enough to know the ID of the sender plus the *public system parameters*. The public system parameters are defined by the private key generator (PKG) during system set-up. The public system parameters include the public key of the PKG and information about the message space. The PKG also generates the private signature keys corresponding to the user IDs.

Figure 10.4 shows a sketch of Shamir's IBS scheme. During the *set-up* phase, the PKG chooses a secret master key and generates the corresponding *public system parameters*.

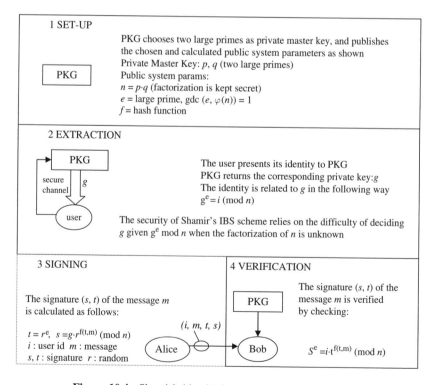

Figure 10.4 Shamir's identity-based signature scheme (IBS)

Afterwards, in the *extraction* phase, it issues *private keys*. The private keys are uniquely given by the IDs and the PKG private master key.

Several IBS schemes have subsequently been proposed. Some examples are found in (Fiat and Shamir, 1987; Cha and Cheon, 2002; Waters, 2005). Boneh and Franklin (2001) introduced the first practical identity-based *encryption* scheme (IBE). This scheme has later been extended by Lynn (2002) to provide message authentication at no additional cost. The cipher text itself serves as the message authentication code.

Integration of identity-based signature and encryption schemes (IBSEs) is studied in Boyen (2003). The latest progress in IBE encompasses strengthened security. Boneh and Boyen (2004) suggested the first IBE scheme proven to also be secure in security models without random oracles. Waters (2005) suggests a more efficient version. However, the IBE, IBSE and IBSC schemes presuppose pairwise communication. None are applicable for network layer one-to-many signing and verification of routing information.

The PKG represents a single point of failure. If the private master key of the PKG is compromised, the entire system is compromised. To counter this, Boneh and Franklin (2001) suggest spreading the PKG master key over more locations using threshold cryptography.

Khalili, Katz and Arbaugh (2003) propose a key management technique (IBC-K) for ad hoc networks combining identity-based cryptography with threshold cryptography (Desmedt, 1994). The nodes that initialize the ad hoc network form a threshold PKG, spreading the PKG private master key over the initial set of nodes by a (k,n) threshold scheme. This eliminates the PKG as a single point of failure and adds intrusion tolerance. It makes the service robust in the sense that an adversary must compromise a minimum of k nodes in order to recover the secret master key. It also reduces vulnerability as the service is available as long as k correctly behaving PKG nodes are within reach.

In order to receive the private key corresponding to some identity, a node must present its identity to k (or more) of the n PKG nodes. The node receives a share of the private key from each of them. With k correct shares, the node can then compute its personal private key.

When time is scarce, physical interaction with a number of geographically distributed PKG nodes is not a good solution. Hence, for scenarios like emergency and rescue operations, a single PKG, e.g. located at the on-site rescue management center, or a hierarchy of PKGs (Boneh *et al.*, 2005) would be more acceptable.

Explicit key revocation remains an unresolved problem. There is no easy way of distributing revocation lists (withdrawing IDs) and making sure all nodes are informed. Another alternative is to change the PKG master key and system parameters. All private keys are derived from these parameters. In essence, an update of the PKG key makes all keys in the system obsolete.

Public Key Schemes – Summary

IBC-K, making certificate exchanges superfluous, is an interesting candidate for ad hoc networks. However, it still relies on a PKG.

10.3.4.2 Symmetric Schemes

Symmetric systems aim to distribute one or more shared secrets through secure channels. Many of the symmetric key management systems for ad hoc networks found in the literature are intended for wireless sensor networks (WSNs). The sensor nodes possess very limited power, memory and computational resources compared to traditional MANET nodes. Symmetric systems may thus be the only option. WSNs normally include a base station. That is, WSNs have a certain amount of infrastructure and are thus not truly ad hoc networks. This survey distinguishes between symmetric schemes for traditional MANETs and WSN schemes. A number of WSN schemes have been included in order to evaluate their applicability in traditional MANETs.

The symmetric key can be distributed either by an online key distribution server or key predistribution. For wireless ad hoc and sensor networks, an online key distribution server is not an option.

A key predistribution scheme consists of three phases: *key predistribution, shared-key discovery* and *path key establishment*. A key server first generates a large number of keys to form a key pool, then every node will be given several different keys selected from the pool and each key will be given a unique identifier. After the nodes are randomly deployed in an area, the shared key discovery phase begins. During this phase, every node attempts to find other sensors in its communication range and will exchange key identifiers. If the node discovers that it shares the same key with its neighbor node, they can use this key for communication. If there is no matching key between the node and its neighbor, it will find a key path to set up a secret communication between them. A key path means that between the two nodes there is a sequence of nodes and each of the two adjacent nodes in the path share a matching key. To establish a secure path with node j, a node i needs to find a path between itself and the node j such that any two adjacent nodes in the path have a common key. Thus, messages from the node i can reach the node j securely.

There are a number of key predistribution approaches. The first is *master key predistribution*. In this solution, all the nodes have a matching master secret key K and any pair of nodes can use this master secret key K to generate a new session key using a specific algorithm. This scheme does not have strong resilience since if one node in the network is compromised by an adversary, the security of the whole network will be damaged. Storing the master key in tamper-resistant hardware can increase the security of the key, but it will increase the cost and power consumption equally.

Pairwise key predistribution can also be used. In this approach, each node in the network carries different keys for every other node in the network so that each pair of nodes N_i and N_j shares a secret pairwise key K_{ij}. This pair of nodes can then use this shared key K_{ij} to generate the session key for their further communication. Node authentication is also available by the pairwise relations. The resilience of this scheme is strong because if one node is compromised by an adversary, the security of other nodes will not be affected. The disadvantage of this scheme is that if the network has a large number of nodes, a large amount of memory is needed for the nodes. In addition, it is difficult to add a new node to the network since the existing nodes do not have the new node's key. Therefore, the scheme does not scale well.

Another approach is *random key predistribution*. In this scheme, each node is randomly given a set of keys from a key pool before it is deployed to the network. Each node may set up connections with other nodes by sharing the same keys or by path key establishment. The key discovery and path key establishment is expected to be successful with probability p. A path key may involve one or more hops. This scheme is scalable and flexible since a new node can easily join the network and it can provide better security when a sensor node is compromised. A few compromised nodes can have only a minor impact when alternative secure paths are available in the network. The disadvantage is that there is a probability that a node may not find a key path with some nodes to set up the secret connection with them. With a probability of $1 - p$ the scheme may suffer connectivity failure when neither a direct shared key nor a path key is found.

Pre-Shared GroupKey (PSGK)

This is an old and well-proven key management scheme with a key distribution center predistributing a symmetric key to all members of the group. A key distribution center could also provide pairwise unique keys, but the focus here is on group keys. The symmetric group key can be used to 'sign' routing information with a cryptographic checksum – MAC (message authentication code).

PSGK lacks intrusion tolerance in the sense that security succumbs to a single captured node. But if the security policy allows it, it is a simple solution. Assuming an offline key distribution center and predistributed keys, the scheme scales well. It is immune to faulty nodes and Byzantine behavior. Authentication should be added offline. With a single group key, there is no easy way to exclude compromised nodes.

PSGK was not designed specially for ad hoc networks. It is included here as several of the symmetric schemes studied basically represent extensions to this scheme.

SKiMPy

SKiMPy (Puzar *et al.*, 2005) was designed for MANETs. It seeks to establish a MANET-wide symmetric key for protection of network layer routing information or application layer user data. On MANET initialization, all nodes generate a random symmetric key and advertise it within one-hop neighborhoods through 'Hello' messages. The *best* key, i.e. the one with the lowest ID number, freshest timestamp or other, is chosen as the local group key. The best key is transferred to the nodes with worse keys through a secure channel established with the aid of predistributed certificates. The procedure is repeated until the 'best' key has been shared with all nodes in the MANET. Once established, the group key serves as proof of trustworthiness. SKiMPy proposes periodical updates of the group key to counter cryptoanalysis. The updated keys are derived from the initial group key.

SKiMPy is bandwidth efficient in the sense that nodes agree on the best key locally. There is no need for an already running routing protocol as the key information is exchanged between neighbors only. SKiMPy implies a delay in spreading the best key to all nodes. Still, the currently best local key can be used to communicate securely until the 'ultimate' key is received.

Byzantine behavior or faulty nodes may disturb local key agreement, e.g. by announcing a better key but not responding.

Entities with special roles or ranks could be empowered to administer certificates. However, online revocation is not possible before the network has been initialized. As the network is initialized, the symmetric group key is also established. Once the symmetric key has been received, there is no efficient way to expel the node from further participation. The group key (or a key derived from it) now serves as proof of trustworthiness. Thus, SKiMPy adds complexity compared to PSGK but does not increase the security accordingly.

Self-Healing Session Key Distribution (S-HEAL)

S-HEAL (Staddon *et al.*, 2002) is a symmetric group key distribution scheme with revocation, designed for networks with unreliable links. The concept demands preshared secrets and a group manager that broadcasts the current group key K 'masked' with a polynomial $h(x)$; $f(x) = h(x) + K$. Individual secrets $h(i)$ are predistributed (i refers to *node ID*). Each member node can then extract the current key by evaluating the received expression at $x = i$ and subtracting the secret value; $f(i) - h(i) = K$. All operations take place in a finite field F_q where q is a prime larger than the number of nodes.

Revocation is enabled by replacing the polynomial $h(x)$ with a bivariate polynomial $s(x, y)$. The group manager now broadcasts the current key K masked as $f(N, x) = s(N, x) + K$. In order to extract the key, the nodes must first recover the polynomial $s(x, i)$ and evaluate it at $s(N, i)$. Then they must subtract the result from the received $s(N, x) + K$, evaluated at $x = i$; $K = f(N, i) - s(N, i)$.

The thinking is that only nonrevoked nodes shall be able to recover the polynomial $s(x, i)$. Given s of degree t, $t + 1$ values are required to find $s(x, i)$. The value N and the individual secrets, $s(i, i)$ are predistributed. The other t values, $s(r_1, x), s(r_2, x) \ldots s(r_t, x)$, that are required to reveal $s(x, i)$, are incorporated in the key update message from the group manager. If the revoked nodes are included in the set $\{r_1, r_2 \ldots r_t\}$, these nodes will only acquire t of the required $t + 1$ values. Consequently, they will not be able to extract the new group key. The scheme enables revocation of a maximum of t nodes.

A main feature of S-HEAL is its *self-healing* property. Nodes that lose one or more key distributions can still reveal the missed keys. Each key update message includes shares of all earlier as well as all possible future keys. The key shares received *before* are complementary to the shares received *after* the key has been distributed. Assuming $p(x)$ is the share received *before* K is distributed, the share received in key update messages *after* K has been distributed equals $K - p(x)$. Hence, missed keys can be derived by combining shares received before the lost update with shares received after the lost update. Whereas the self-healing feature may be of great value in mail systems and similar applications, network layer routing information has only instant value. Hence, retrieving earlier keys is of little interest. Further details are therefore left out.

S-HEAL's reliance on a group manager – possibly multiple hops away – to provide the initial group key makes it inapplicable for protection of routing information. The group key is needed in order to bootstrap the network service but S-HEAL demands an already running network service to distribute the group key. Nevertheless, S-HEAL could potentially be used for revocation and rekeying, assuming a protected network service has been bootstrapped with an initial predistributed group key (PSGK). This would improve intrusion tolerance compared to pure PSGK. Robustness to packet losses could be increased by periodically retransmitting the latest key update rather than waiting for the next key update as implied by (Staddon *et al.*, 2002).

Regarding scalability, the message sizes and number of key update messages are independent of the number of nodes in the network. The size of the key updates is only proportional to the size of the polynomials (if self-healing is left out.)

Missing source authentication of the broadcasts from the group manager is a shortcoming. A MAC generated by the previous group key could easily be added. Still, a Byzantine behaving node could potentially transmit garble, claiming to be the next key from the group manager, and cause disruption.

Logical Key Hierarchy (LKH)

Group keys can be updated by brute force: a group manager distributes the new group key, encrypted with a separate (individual) key for each node. In essence, LKH represents a family of schemes that improve the scalability of this brute force method by organizing the keys into a logical hierarchy and giving the nodes additional keys.

LKH was introduced by Wong, Gouda and Lam (1998) and Wallner, Harder and Agee (1999). The concept is illustrated in Figure 10.5. All group members (N1–N8) possess the group key $K_{12345678}$. The sub-group key K_{1234} is shared by members N1–N4, and K_{12} is common to N1 and N2. $K_1 - K_8$ refer to the individual keys. Assuming node N8 is to be revoked, all group and sub-group keys known to N8 ($K_{12345678}$, K_{5678} and K_{78}) should be updated. N7 shares all intermediate keys from the leaf to the root with N8. N7 must therefore receive the updated keys encrypted with its individual key. The new group key and sub-group key can be distributed to N5 and N6 encrypted with their key in common – K_{56}. To N1–N4, the group manager sends the new group key $K_{1234567}$ encrypted with K_{1234}. Thus, bandwidth and computational cost is saved compared to updates encrypted with the individual keys.

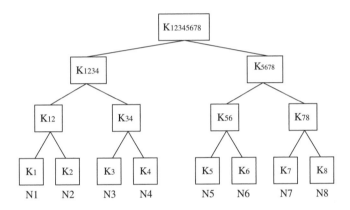

Figure 10.5   Logical key hierarchy

The key tree may be binary or *k*-ary, and balanced or unbalanced.

Whereas the basic LKH scheme was not designed specifically for ad hoc networks, Rhee, Park and Tsudik (2004, 2005) suggest an LKH scheme for hierarchical ad hoc networks. They propose that the group manager functionality is distributed over several managers, each controlling different *cells* of the network. The approach is cellular and infrastructure-based rather than ad hoc. The nodes are fully dependent upon the cell managers. Each cell has a different group key.

The nodes must contact the cell manager to receive the key when they move from one cell to another. In other words, the nodes must be able to detect cell boundaries and be within communication range of a cell manager. In addition, the scheme requires that cell managers communicate during 'key hand-off' from one cell to another. The intention of the scheme is to limit rekeying to part of the network. The price is reduced robustness and increased bandwidth consumption. The scheme is inapplicable for MANET use.

A number of other refinements to the basic LKH scheme (Wong *et al.*, 1998; Wallner *et al.*, 1999), focusing on communicational and computational cost, have been proposed. OFT (Balenson *et al.*, 2000), OFC (Canetti *et al.*, 1999), ELK (Perrig *et al.*, 2001a), LKH+

(Balenson *et al.*, 2000), EBHT (Rafaeli *et al.*, 2001), LKH++ (Pietro *et al.*, 2002), Pooven-dran and Baras (1999) and the Internet draft by Selcuk, McCubbin and Sidhu (2000) all propose different ways to reduce communication overhead – primarily focusing on message sizes. Of these schemes, only LKH++ is designed for wireless networks.

ELK reduces the size of the key update messages by sending only part of the key plus a key verification value. The receivers must search by brute force for the remaining part of the key. The verification value is used to decide whether the correct key has been found or not. LKH+ and EBHT suggest new keys are derived by applying a one-way function to the old key(s) when members are added. In OFT, OFC and LKH++ the keys of parent nodes are related to keys of their child nodes through one-way functions. After a group change, the group manager only sends enough information to enable the nodes to compute the rest of the updated keys themselves. Poovendran and Baras (1999) show that the overhead can be reduced by placing nodes that are most likely to be revoked as close as possible to the root, i.e. giving them only a minimum number of keys. Similar ideas are also studied in Selcuk et al. (2000). In a practical situation it can be hard to decide which nodes are most likely to be compromised. Furthermore, in ad hoc networks, the *number* of messages may be more devastating than the *size* of the messages (Winjum *et al.*, 2005). Hence, the actual gain of the proposed refinements is not evident. Simulations are required in order to judge which approach is the best.

LKH++ claims to reduce the number of messages, as some of the nodes will be able to calculate the new key by themselves. The keys of parent nodes are related to keys of their child nodes through one-way functions. According to LKH++ and referring to Figure 10.5, K_{12} equals a hash of K_1, and K_{1234} represents a hash of K_{12}. Consequently, children to the left will be able to calculate the new key of their parent node. The others receive it from the group manager. Still, the nodes to the left must also be made aware that a key update is required. LKH++ does not address how.

In an ad hoc network, rekeying every time a new node joins or leaves the network is unnec-essary and undesirable. Routing information has only instant value. Backward and forward secrecy on joining and leaving nodes is of little importance. However, LKH may be of inter-est as an extension to PSGK for revocation purposes. Assuming the network service has been initialized (with the aid of the predistributed group key), LKH could be used to expel compro-mised nodes. A static tree, large enough to hold the keys of all anticipated members, would be required in order to avoid rekeying when new nodes are added.

For (infrequent) revocations in ad hoc networks for emergency and rescue operations, robustness is even more important than communicational and computational cost. In the basic LKH scheme, innocent nodes that miss the update from the group manager may be cut off. Periodical retransmissions of the last update(s) could help. In ELK, the group manager sends repeated hint messages that enable nodes that lost the key update to calculate the key. Forward error correction codes (FECs) on key updates, as suggested by Wong and Lam (2000), enable correction of bit errors but do not help nodes that missed the entire update.

The group manager represents a single point of failure. Replication of the group manager for reliability and performance, as suggested by Wong *et al.* (1998), is of limited value for ad hoc networks. Replication demands synchronized servers and increases the number of targets for security attacks.

A general weakness of schemes that rely on symmetric keys only is that there is no possi-bility for source authentication. A Byzantine node may pose as the group manager and cause disruption. Wong *et al.* (1998) propose authentication through digital signatures. The basic key management problem then reverts to public key distribution.

Probabilistic Key Predistribution (PRE)

PRE (Eschenauer and Gligor, 2002) assumes WSN nodes outfitted with a preinstalled *key ring*, i.e. a set of keys drawn randomly from a large pool of keys. When bootstrapping the network, the nodes broadcast the identifiers of the keys in their key ring. A wireless link is established between nodes only if they share a key. Hence, resilience to Byzantine behavior and faulty nodes is fine. The scheme relies on a controller node (base station) to broadcast a signed list of the key identifiers to be revoked.

A number of probabilistic key predistribution schemes for WSNs have been proposed. Chan, Perrig and Song (2003) suggest extensions to the work of Eschenauer and Gligor (2002) that increase the resilience to node capture. This requires q common keys ($q > 1$) instead of just a single one to establish a connection. Liu and Ning (2003a) propose probabilistic preshared *polynomials* for the establishment of pairwise keys in WSNs. Polynomial sharing increases resilience to captured nodes. A trusted entity defines a bivariate polynomial, $f(x, y)$ with the property $f(x, y) = f(y, x)$. Secret polynomial shares, $f(i, y)$, are predistributed to each sensor node, i. Any two nodes, i and j, can set up a pairwise key by evaluating the polynomial at $f(i,j)$ and $f(j,i)$, respectively. Similarly, Du, Deng, Han and Varshney (2003) suggest another scheme relying on probabilistic preshared polynomials for pairwise keys in sensor networks. Du *et al.* (2004) suggest the use of deployment knowledge to increase the probability that two nodes find a common secret key. The latter may be possible in a WSN with planned positioning of sensors but not in a MANET.

Zhu *et al.* (2003a) propose probabilistic key predistribution combined with secret-sharing to set up pairwise exclusive keys in MANETs. A node wishing to communicate securely with another picks a secret symmetric key. It then sends shares of this secret symmetric key, encrypted with different predistributed keys, to the opposite party, i.e. the shares are sent through different logical paths. Assuming the aggregated set of predistributed keys used is known to the two nodes in question only, no other nodes will be able to decrypt enough shares to reveal the secret symmetric key. Depending on configuration, the scheme may produce a large number of messages. Zhu *et al.* (2003a) claim it is desirable to trade computation for communication in ad hoc networks. This assumption does not generally hold.

The objective of the key ring in PRE is intrusion tolerance. The price is availability. There is only a probabilistic assurance that a node actually will share a key with one or more neighbors and be able to bootstrap communication. Emergency and rescue operations, where availability is a number one concern, would require a key ring large enough to achieve close to zero probability of failure. The consequence is intrusion resistance reduced to a level comparable to a predistributed group key (PSGK). This contradicts the intention of the scheme. The applicability and scalability for network layer security is limited. Different keys in common with the various neighbors imply more signatures for each routing message. End-to-end signatures on routing messages to be flooded are precluded.

Security Protocols for Sensor Networks (SPINS)

The SPINS security protocols for WSNs (Perrig *et al.*, 2002) assume preinstalled individual (pairwise) keys between the sensor nodes and a base station. Nodes that want to communicate securely request a common key from the base station. The base station returns the key, encrypted with their individual keys. This scheme demands an already running routing protocol and reliable access to the base station. It is inapplicable for the purpose of protecting routing information in a traditional MANET.

SPINS also includes a scheme for authenticated broadcast – $\mu TESLA$ – and describes how this can be used to provide an authenticated routing protocol for sensor networks. μTESLA relies on a predistributed *commitment*, i.e. the last key of a one-way key chain, and delayed disclosure of subsequent keys in the key chain. The key chain can be derived by repeated hashes of an initial random key. The key used at time i equals a hash (or similar one-way function) of the key used at time $i + 1$. The commitment enables the nodes to verify that later disclosed keys originate from the claimed source; repeated hashes of the disclosed key should return the commitment. To send an authenticated packet, the sender computes a message authentication code (MAC) with a key that is secret at that point in time. The receiver stores the message until the key is later disclosed. The nodes must be loosely time synchronized and know the key disclosure schedule. Otherwise, adversaries could forge packets as the receiver would not know whether the key used to calculate the MAC of an incoming packet had been disclosed or not.

The SPINS authenticated routing protocol discovers routes from the nodes to the base station with the aid of μTESLA key disclosure packets flooded from the base station. The sender, from which a node first received the valid μTESLA packet, is set as parent node in the route to the base station. The predistributed commitment enables the nodes to verify that the received packet originated from the base station.

This may work for communication from sensor nodes to a base station. The same technique cannot be used in a traditional MANET with a scattered communication pattern. One possibility would be to preload all nodes with commitments of the key chains of all other nodes. This would allow any node to authenticate the messages from any other node. Intrusion tolerance would be fine, and robustness to Byzantine behavior and fault nodes would be good. However, the nodes would have to be loosely time synchronized and know the key disclosure schedule of all other nodes. The solution would give little flexibility and scale badly. Furthermore, delayed key disclosure is problematic in the setting of mobile nodes and rapidly changing network topology. Altogether, the SPINS key management scheme and authenticated routing protocol is inapplicable for protection of routing information in traditional MANETs.

GKMPAN

GKMPAN (Zhu *et al.*, 2004) was designed for secure multicast in ad hoc networks. It is basically a revocation and rekeying scheme for PSGK, founded on PRE (Eschenauer and Gligor, 2002) and μTESLA (Perrig *et al.*, 2002). GKMPAN assumes a predistributed *group key* plus a predistributed *commitment*. The group key is used to protect multicast communication. The commitment is used for authentication of revocation messages from the *key server*. In addition, GKMPAN assumes each node is equipped with a preinstalled subset of symmetric keys drawn from a large key pool.

In contrast to PRE, the keys in the key set are determined from the ID of the node. On revocation, the key server issues a revocation message containing the ID of the revoked node. All keys in the key set of the revoked node should be erased or updated. Any node can automatically tell from the ID which keys to revoke. The revocation message also identifies a key that is not in the key set of the revoked node, to be used as the 'update key.'

A new group key is derived from the old one with the aid of a keyed one-way function; the old group key is used as data input and the 'update key' as key input. The output is the new group key. Nodes that have the 'update key' in their key set can calculate the new group key without assistance. The others receive it from their parent node, encrypted with one of the (nonrevoked) keys in their key set. It is distributed through a multicast tree rooted at the key server. The validity of the revocation message cannot be checked before the key server later discloses the key that was used to compute the message authentication code.

In order to avoid potential disruptions, the old and new group key should co-exist until all nodes have received the new group key. However, there is no easy way to make sure that all nodes have received the new group key. Byzantine behavior or faulty nodes may inhibit efficient exclusion and rekeying. The reliance on a key server and time synchronization are other vulnerabilities. GKMPAN scales fairly in the sense that new group keys can be calculated by the nodes themselves or transferred locally. In addition, GKMPAN increases intrusion tolerance compared to PSGK as it enables node exclusion. The price is reduced availability. Innocent nodes may be expelled if all their keys happen to be in revoked key sets. This is not acceptable in settings like emergency and rescue operations.

Secure Pebblenets (PEBL)

Pebblenets (Basagni *et al.*, 2001) refer to large ad hoc networks where the nodes are called *pebbles* due to their small size and large number, e.g. WSNs. The aim of PEBL is to protect application data. It establishes and updates a network-wide traffic encryption key, TEK. At the network layer a preinstalled group key guarantees the authenticity of a pebble as a member of a group. Hence, PEBL can be regarded as an extension to PSGK. The assumption is that only nodes possessing the group key are capable of encrypting and decrypting 'Hello' messages correctly. Furthermore, PEBL assumes the pebbles organize into clusters of one-hop neighbors. Each cluster selects a cluster head node. The cluster heads establish a backbone and compete to become the *key manager*. The key manager generates the traffic key, TEK, which is intended for encryption of application layer data traffic. The TEK is distributed from the key manager to the regular nodes through the cluster heads. It is updated periodically. Each TEK update is preceded by a reclustering and new selection of cluster heads. This rotation of the cluster head role is to avoid exhaustion of the nodes acting as cluster heads, and to account for mobility. Nodes that were one-hop neighbors when the cluster was formed may have moved out of the neighborhood.

Nodes that do not behave according to the protocol may disturb cluster formation and TEK updates. PEBL offers no protection against replay or intrusion. PEBL security succumbs to tampering. Both network layer 'Hello' messages and TEKs are all protected by keys derived from the group key. Anyone possessing the group key will be able to participate in the TEK updates. PEBL in its entirety, with cluster formation and periodic TEK updates, is bandwidth consuming, demands synchronization and makes availability assumptions that render it unsuitable for MANET use.

Key Infection (INF)

INF (Anderson *et al.*, 2004) is intended for WSNs. The scheme assumes static sensor nodes and mass deployment. INF sets up symmetric keys between the nodes and their one-hop neighbors. The security is based on surprise: it relies on the assumption that during the network deployment phase, any attacker is only able to monitor a fixed percentage of the communication channels. At bootstrap time, every node simply generates a symmetric key and sends it in the clear to its neighbors. A *key whispering* approach is used, i.e. the key is initially transmitted at a low power level. The transmission power is then increased until the key is heard by at least one of its one-hop neighbors and a reply is received. INF is simple, self-organizing and robust to Byzantine behavior and faulty nodes. It is bandwidth efficient and scales well. However, the security is weak. INF is vulnerable to eavesdropping during key whispering. In addition, there is no authentication of the communicating parties. INF's 'security through surprise' fails for MANETs where static nodes and instant mass deployment are not options.

Localized Encryption and Authentication Protocol (LEAP)

LEAP (Zhu *et al.*, 2003b) was designed for static WSNs. LEAP suggests different keys for different purposes. It requires a number of predistributed keys. Predistributed *individual keys* are used for communication between sensor nodes and the base station. A preshared *group key* is applied for protection of broadcast information from the base station. A preinstalled network wide *initial key, K*, is used to derive *pairwise keys* for secure communication between one-hop neighbors.

During neighbor discovery immediately after deployment, each node n derives its master key, K_n. The master key is derived as a function of its node ID and the initial key; $K_n = f_K(ID_n)$. The master key is used to 'sign' 'Hello' messages. Any node that knows the initial key is able to calculate the master key of any other node ID. Hence, each node can verify the 'Hello' messages received from its neighbors. The node then calculates the pairwise keys shared with its neighbors, v, as a function of their master key and the node ID $- K_{nv} = f_{Kv}(ID_n)$.

Intrusion tolerance is obtained under the assumption of stationary nodes; the network key is erased after the pairwise keys have been established. Nodes that have erased the network key can no longer establish pairwise keys. New nodes can still be added though. As the new nodes have not yet erased the group key, they can set up pairwise keys with their neighbors. When a node is captured, only the keys held by the captured node are compromised.

The pairwise keys are used both to secure ordinary data and to distribute *cluster keys*. The cluster keys are employed for secure local broadcasts. Any node simply generates a cluster key and sends it to all neighbors, encrypted with the respective pairwise keys.

Whereas LEAP may work in a static sensor network, the heart of this key management scheme – the setting up of pairwise keys – will not work in a traditional ad hoc network. Deletion of the initial key is incompatible with mobile nodes and constantly changing network topology. Evaluating the scalability of LEAP in MANETs makes little sense, as pairwise key set-up is precluded after the initial key has been erased.

Distributive Symmetric Schemes: Summary

An overview of the capabilities of the distributive symmetric key management schemes has been given. The WSN key management schemes generally assume static nodes, mass deployment, node-to-base station communication patterns or the establishment of pairwise keys. Their aim and assumptions render them inapplicable for protection of routing information in traditional ad hoc networks with mobile nodes. PSGK or PSGK extended by S-HEAL or LKH for revocation appear to be the most promising alternatives to the symmetric schemes.

10.4 Authentication Issues

During data communication, an adversary can easily modify the data and inject some messages into the data, so the receiver should make sure that the data received come from a legitimate sender and have not been tampered with. Data authentication allows a receiver to verify that the data really were sent by the claimed sender. So the receiver needs to ensure that the data used in any decision-making process originate from the correct source. In addition, authentication is necessary for many administrative tasks in network construction. Data authentication is also important in sensor networks.

In the two-party communication case, the sender can create a message authentication code (MAC) by computing a checksum with the message content using a secret key. Data authentication can be verified by the receiver that has both the shared secret key and the original message content that was used to create the same MAC. However, in multiparty communication, such as a base station broadcasting data to a number of nodes, this symmetric data authentication cannot be used. In this case, an asymmetric mechanism, such as TESLA, will be adopted. In this approach, first the sender will broadcast a message with a MAC that has been generated with a secret key, which will be disclosed later. When the node receives this message, it will first buffer the message if it has not received the key disclosed by the sender. After it receives the key, it will use this key together with the buffered message to generate the MAC to authenticate the message. The disadvantage of TESLA is the initial parameters for the authentication should be unicast to every receiver, which is not efficient for a network that has a large number of nodes. In Liu and Ning (2003b), a multilevel key chain is used for the key distribution and the initial parameters are predetermined and broadcast to the receivers other than by unicast. By doing so, this increases the scalability of a network with a large number of nodes and at the same time resists replay and denial-of-service attacks.

10.5 Integrity

Data integrity means the data received by the receiver are the same as the data generated by the sender. In wireless sensor networks, if a node is captured by an adversary, the adversary may modify the data or inject some wrong message into the network. Also, because of the limited resources of the node and the harsh environments where the node is deployed, communication data may be lost or damaged or the data integrity may be destroyed. To protect data integrity, the simplest mechanism is to use the cyclic redundancy check (CRC); another way is to use encryption-based integrity methods such as a message authentication code (MAC) as in authentication, which is stronger but more complicated.

Confidentiality may prevent information disclosure. However, the data may still be modified in order to disrupt the communication. For instance, a malicious node may add some fragments or manipulate the data within a packet. This new packet can then be sent to the original receiver. Data loss or damage can even occur without the presence of a malicious node due to the harsh communication environment. Thus, data integrity ensures that any received data have not been altered in transit.

10.6 Review Questions

10.1 What are the challenges associated with the security of wireless ad hoc networks and sensor networks?

10.2 What is the general procedure to establish a secure network?

10.3 Compare the bootstrapping requirements for wireless ad hoc networks and wireless sensor networks.

10.4 Outline the different approaches to key predistribution.

10.5 What are the advantages and disadvantages of using a base station as the online key distribution server?

10.6 What are data authentication and data integrity in network security?

11

Challenges and Solutions: Protection

11.1 Privacy and Anonymity

The presence of wireless ad hoc and sensor networks all around us exacerbates privacy issues that must be addressed in order to prevent unauthorized observers. With the ubiquitous deployment of sensor nodes, there is now higher potential for abuse of the information collected. In contrast to direct site surveillance, information can now be made available through remote access to wireless sensor networks, and multiple sites can be monitored simultaneously by a single adversary, who can do so in a low-risk, anonymous manner. In addition, a major challenge comes from the possibility of correlating datasets to conclude new information. If the right method is used, sensitive information may be derived from even seemingly innocuous data; for instance, location information from sensors can enable the possible identification of a user, making continuous movement tracking feasible. All of this aggravates the privacy problem (Chan and Perrig, 2003).

There is obvious conflict between the need for public information and the demand for personal privacy. For example, the location of an event detected by a sensor may be required by certain applications while the location of a user constitutes a privacy issue. Total anonymity is difficult in many cases and compromises often have to be reached.

Anonymity techniques are needed to prevent an adversary from identifying the sender and receiver. To achieve anonymity, data should be depersonalized before release. There are four well-known approaches to anonymity (Priyantha *et al.*, 2000; Smailagic *et al.*, 2001; Gruteser *et al.*, 2003):

- Sensitive data can be decentralized, e.g. distributed through a spanning tree, so that a complete view of the original data requires collaboration from a set of distributed nodes.
- Eavesdropping and active attacks can be prevented by using secure communication protocols, such as SPINS.
- To protect against traffic analysis, data traffic can be changed by simple de-patterning of the data transmission; for example. padding with some bogus but real-looking random data can

---

Security in Wireless Ad Hoc and Sensor Networks   Erdal Çayırcı and Chunming Rong
© 2009 John Wiley & Sons, Ltd

intensively change the traffic pattern when necessary. Information flooding will be further discussed later in this section.

- In relation to location sensing, increasing sensor node mobility can be an effective defense-of-privacy method. Location sensors can be placed on the mobile device instead of in the monitoring infrastructure. Physical location can be determined, for instance, by using passive listeners that hear and analyze information from beacons spread throughout the area. The location information is now in the hands of the user and the user can choose the parties to which the information should be transmitted.

As mentioned above, information flooding may be an efficient way to provide anonymity and solve the privacy problem in sensor networks. Flooding-based algorithms that provide additional privacy and anonymity are explained in detail in Section 12.1.5.

Finally, policy-based access control decisions and authentication can also be used to address the privacy problem. The privacy policies in various domains may be specified based on criteria such as identity, time and location. For instance, access control to a centralized location server from clients can be provided through validating a set of XML-encoded application privacy policies.

11.2 Intrusion Detection

Wireless ad hoc networking is associated with vulnerable characteristics such as open-air transmission and self-organizing without a fixed infrastructure or centralized management. Consequently, ad hoc networks are more susceptible to attack, and the security challenges in them are more complicated. As the first line of defense, intrusion-prevention techniques, such as encryption and authentication, can be used to defend against intruders. However, even in a fixed-wire network, proactive defense alone is not sufficient to secure a system from all penetrations. A second line of defense system is needed to detect an ongoing attack in the network. If such detection is available, damage may be minimized.

An intrusion-detection system (IDS) monitors activities in a system and then analyzes the audit data to determine whether there is a violation of the security rules. An alert is given if a violation known to be malicious is found. Responses to the attack may also be initiated by the IDS accordingly. The available techniques include abnormality, misuse and specification-based detection (Mishra *et al.*, 2004):

- To detect abnormality, normal behavior profiles of users are kept in the system. Any system activity that deviates from the baselines in a corresponding profile is considered a possible intrusion. For this approach it is essential that normal behavior profiles be updated periodically.
- To detect misuse, the system uses patterns of well-known attacks kept in the system to match and identify known intrusions. However, it cannot detect new types of attack.
- In specification-based detection, a set of constraints is predefined to describe the correct operation of a program or protocol. Any activity deviating from these constraints is treated as a possible intrusion.

11.2.1 Architectures for IDS in Wireless Ad Hoc Networks

The nature of wireless ad hoc networks makes them very vulnerable to attack. First of all, the mobile nodes are independent and their movements are not controlled by the system, so they can easily be captured, compromised and hijacked. Secondly, since in wireless networks there are no physical obstacles for the adversary, attacks can come from all directions and target any node. Third, in wireless ad hoc networks adversaries can exploit the decentralized management for new types of attack designed to break the cooperative algorithms. To tackle these additional challenges, several possible IDS architectures exist including standalone IDS, distributed and cooperative IDS and hierarchical IDS.

11.2.1.1 Standalone IDS

Each node has its own IDS and detects attacks independently in this architecture. There is no cooperation between nodes and all decisions are based on information collected by individual nodes. This architecture is not effective enough but can also be utilized in networks where not every node is capable of running IDS.

11.2.1.2 Distributed and Cooperative IDS

Since wireless ad hoc networks are distributed and based on cooperation among nodes, the intrusion detection and response system should also naturally be distributed and cooperative. In this architecture (Zhang *et al.*, 2003), each node has an IDS agent and makes local detection decisions by itself. At the same time, all the nodes participate in a global detection process. Like the standalone IDS architecture, the distributed and cooperative IDS structure is more suitable for a flat network configuration than a cluster-based multilayered one.

11.2.1.3 Hierarchical IDS

In multilayer wireless ad hoc networks nodes are divided into clusters. To fit the requirements of multilayer wireless ad hoc networks, hierarchical intrusion-detection systems are proposed, where each node has its own IDS agent responsible for local intrusion detection. At the same time, the IDS agent of the cluster head is responsible for both local and global intrusion detection. Total network coverage is assured by activating global agents in every cluster head. However, the clustering also adds possible points of attack and overhead and complexity in the creation and maintenance of clusters.

In Roman *et al.* (2006), an alternative distributed solution, called *spontaneous watchdogs*, is suggested for flat sensor network architectures without organizing them into clusters or adding more powerful nodes. Some sensor nodes are chosen independently as spontaneous watchdogs to monitor the communications in their neighborhoods. The technique relies on the broadcast nature of sensor communications and takes advantage of the high density of sensor nodes being deployed in the field. Each circulating packet may be received by a set of sensor nodes either in one broadcasting range or in the next-hop as a relayed packet. Hence, all these sensor nodes have a chance to activate their global agents in order to monitor these packets.

11.2.1.4 Mobile Agent for IDS

A mobile-agent-based IDS can be considered either a distributed and cooperative intrusion detection technique or it can be used in combination with hierarchical IDS. An agent is mobile due to its ability to move through the network and interact with nodes and collect information from them. The intrusion-detection tasks are distributed and assigned to these mobile agents. Each mobile agent is assigned a specific task and acts upon the information it collects along its moving path.

There are many advantages (Mishra *et al.*, 2004) of using mobile agents. First of all, power consumption of the network is reduced because the tasks are distributed and each node holds only some of the tasks and not all of them. Secondly, the overall system fault tolerance increases because the IDS tasks are distributed to different parts of the network; when some agents are destroyed or parts of the network are separated, the other agents can remain functional. Third, as the mobile agent may be platform independent, the IDS can run under different operating system environments. Furthermore, when a central processing unit is replaced by distributed mobile agents, the computational load is divided between machines and the network load is reduced. However, these mobile agents still need to be run in a secure module on each node in order to protect themselves on remote hosts.

11.2.1.5 IDS for Sensor Networks

Wireless sensor networks share many similarities with wireless ad hoc networks, but there are also many differences between them. In wireless sensor networks, tiny and simple sensor nodes are constrained in terms of energy supply and computational capability. Due to their limited range, the node density in wireless sensor networks is usually higher than in ad hoc networks. In wireless sensor networks, most sensors are stationary and lack mobility. These constraints, inherent to wireless sensor networks, together with sensor network architectures make it difficult to design an IDS for wireless sensor networks.

Although the above IDS solutions for ad hoc networks can be applied to sensor networks (Zhang *et al.*, 2003), it may be inappropriate to use them directly. For instance, due to the power constraint, it may not be feasible to have a full-powered IDS agent in every node or to force every node to analyze all the packets from its neighbors in a high-density sensor network area. How to distribute the detection tasks over the nodes is the most fundamental problem associated with IDS solutions in sensor networks.

An IDS system for wireless sensor networks must meet the following challenges (Roman *et al.*, 2005):

1. A simple and highly specialized IDS architecture is needed to analyze the specific network protocols and react to the specific threats in sensor networks. An IDS for a sensor network should utilize as few resources as possible in parsing its set of detecting rules and in communications between nodes for information exchange. For instance, it is vital to manage memory space. Scanning results from IDS must consume little memory space, and the policy must be able to continue working when the memory is full.

2. Once any agent discovers a possible breach of security in the network, it must create and send an alert to the base station quickly.

3. The IDS system must be distributed and interchange information to achieve better performance; this means that the data collection and analysis will be performed in different locations.
4. The IDS should be able to withstand hostility.

11.3 Defense Against Traffic Analysis

Wireless sensor networks consist of many small, resource-constrained sensor nodes together with several base stations. The base stations are often more powerful and have larger computational abilities than the ordinary sensor nodes. They play an important role in the system and hence become the main targets for attack.

If an adversary can successfully attack the base station, the entire system can be rendered useless easily. Therefore, locating a base station would be very helpful for an adversary. Adversaries can locate base stations and obtain important information about the network simply by monitoring the traffic pattern and traffic volume, even when the packets are encrypted. In sensor networks, the data collected by sensor nodes are typically routed to base stations through relatively fixed paths, and the nodes near the base station forward far more packets than the nodes far away from the base station. Based on these characteristics, adversaries can analyze the traffic patterns and reveal the location of the base station without understanding the contents of the packets. This is known as a *rate monitoring attack* (Deng *et al.*, 2005). In order to deceive and misdirect an adversary, packets collected by the sensor nodes may be forwarded randomly other than to the parent node, in the hope that the adversary will not easily find the exact path to the base station.

In addition, there is the *time correlation attack*. In this attack (Deng *et al.*, 2005), adversaries can deduce the location of the base station by simply generating some events and monitoring where the sensor nodes forward the packets. One way to defend against the time correlation attack is to buffer incoming packets in the node for a random time before forwarding them. Another way is to generate a fake packet with some probability and send it randomly to another node every time a node wants to send a packet. The fake packets use a time-to-live (TTL) to decide when forwarding should stop.

11.4 Access Control and Secure Human–Computer Interaction

Many security failures have their sources in human error and human involvement. In some cases even the most secure system may fail if it is used incorrectly or bypassed by users. However, simply labeling users the 'weakest link in the security chain' will not give us a more secure system. It is important to consider human–computer interaction (HCI) when designing a security system. A user-unfriendly system can lead to serious errors and consequences related to security by its users. Previous HCI knowledge and techniques can be used to prevent or address known problems and to design a more user-friendly and secure system.

Access control using passwords is a good HCI example when considered in relation to user behavior. Access control refers to the practice of gaining entry to restricted areas by authorized personnel. Access control to a computer system involves identification, authorization and auditing. Identification and authentication are the two steps in the service that determine who is allowed to enter a system. Identification is where a user tells the system who he or she is, based on a relatively simple mechanism like a username or user id. Authentication is the process of verifying a user's claimed identity. By comparing an entered password to the password stored in a system for a given username, the user can access the system if a match is found, and be refused otherwise. If a user fails to provide a match for three or five times in sequence, the account may be suspended and may not be used until it has been reset by contacting the system administrator.

However, there are many problems related to the use of a password mechanism. First of all, since every system has its own password mechanism, when a user wants to log in to different systems he needs to enter the username and password several times. An obvious solution to this is for the systems to store the usernames and passwords automatically, so that the users do not need to input them again, but security in this solution is not good. Second, according to password policies, the passwords selected by users need to satisfy the following requirements: passwords must be strong, i.e. a pseudo-random mixture of letters, numbers and characters; users should have a different password for each system; passwords should be changed at regular intervals and accounts of users who do not comply should be deleted or suspended. This means more human memory is needed to deal with the service. Third, since the requirements for usernames and passwords for different systems vary greatly, the users need to remember far more. They must remember the passwords and usernames for different systems and the password restrictions of each system.

Digital access is required increasingly widely in our daily lives and our places of work. Consequently, there are more and more usernames and their corresponding passwords for each of us to remember. It is very difficult for most people to remember them all exactly. Users are increasingly unable to cope with the proliferation of passwords. If a password is forgotten, resetting the password is needed and this is not without cost. Many normal users choose the easy solution: write down the usernames and related passwords of different systems and keep them in a 'safe place' (e.g. in their desks). This violates the principle of knowledge-based authentication – that the password should be kept only in the system and the user's mind.

For the security of systems, human memory must be studied and the findings incorporated into password design. The following important characteristics should be recognized (Sasse *et al.*, 2001):

- people may not recall an item 100% correctly;
- recognition of a familiar item is easier than unaided recall;
- frequently recalled items are easier to remember than infrequently used ones, and retrieval of very frequently recalled items becomes 'automatic';
- people cannot 'forget on demand' – items will linger in the memory even after they are no longer needed;
- items that are meaningful are easier to recall than nonmeaningful ones;
- distinct items can be associated with each other to facilitate recall and similar items compete against each other on recall.

Single sign-on (SSO) is a method of access control that enables a user to authenticate once and gain access to the resources from multiple systems. This can reduce both the number of passwords a user needs to remember and the number of logons. However, SSO requires a homogeneous infrastructure, or at least a unified user entity authentication scheme together with a centralized user database.

If SSO is not available, using only one username and password for different systems is another approach. At first, this approach does not seem to be a good choice because all the systems will be compromised if this single password falls into the wrong hands. However, from the usability point of view, this approach is desirable because it reduces the users' memory load. Hence, a user can choose a stronger password, which in turn can bring more security to the system.

The problem with strong passwords is that they can be difficult to remember. A good solution is to create passwords by adding a combination of letters, numbers and characters that are meaningful to the user. This can make them easy to remember and at the same time difficult for adversaries to crack. Experiments show that, for ordinary users, generating passwords by concatenating several words with some characters is a good choice. For heavily used passwords, many system administrators generate passwords from a sentence. For example 'MbCg40K!' can be generated from 'My boat can go 40 knots.'

Several kinds of authentication mechanism that use cued recall have been developed to help people remember passwords more easily. Composite weak authentication, cognitive passwords and associative passwords are some of these password schemes (Sasse *et al.*, 2001). For composite weak authentication, customers can identify themselves by providing several memory items for authentication. Cognitive passwords are based on personal facts, interests and opinions that are likely to be easily recalled by a user. When a user provides a system with exact answers to a rotating set of questions, he passes the authentication process. For associative passwords, a word pair that has some association is used to form passwords.

Alternatively, there are systems that use visual memory as a means of user authentication. In these systems, instead of recalling a password, people need to recall an image seen previously. In general, it is easier for people to recognize something than to recall the same information from memory without help. Recognition-based authentication systems can achieve better security than systems that rely on recall-based authentication.

In conclusion, there are several approaches to the easy implementation of password-based access control systems. First, to provide a unified mechanism to manage the authentication of users, single sign-on (SSO) can be used. With the SSO mechanism, a single action of user authentication and authorization can permit a user to access all computers and systems that he/she can access. By reducing the number of enforced changes and using the same password for several systems, password inflation can be reduced. This is safer than writing down passwords to remember. For an infrequently used password, a two-step procedure may be a good choice since it affords a chance to give help in recalling a password. Finally, user knowledge and motivation should also be taken into account when designing password mechanisms.

Some may suggest that knowledge-based authentication should be replaced by biometric authentication. However, this is unlikely to happen for a while because biometric-based authentication still does not work well for all security systems.

11.5 Software-Based Anti-Tamper Techniques

Currently, software cracking is a big problem for the software industry. To protect software from being cracked, both hardware-based and software-based anti-tamper technologies have been developed. Software protection has recently attracted tremendous attention and more and more anti-tamper technologies are being proposed for this purpose.

Anti-tamper techniques in general are designed to detect or sense any type of unauthorized modification or use of software. Once such tampering is detected, the anti-tamper part of the software will take some action to render the software useless to the adversary.

Software cracking is the modification of an application's executable code to cause or prevent a specific key branch in the program's execution. An ordinary example of software cracking is removing the expiration period from a time-limited trial of an application. Several kinds of attack fall under the banner of software cracking, including gaining unauthorized access, reverse engineering and violating code integrity. To gain unauthorized access, an attacker can access the software by disabling its access control mechanism. If the attacker can gain unauthorized access to the software, he can make an illegal copy or modify some functions of it. Reverse engineering can reverse a program's machine code back into the source code in which it was written by reversing the complied program code using a debugger. The goals are to study how the program performs certain operations in order to remove protection from the software, or to reuse part of the code in other software. Violating code integrity means the injection of malicious code into the program. The execution of the malicious code allows the attacker to gain privileges on the executing program.

To combat cracking, a wide range of anti-tamper protection mechanisms have been studied (Atallah *et al.*, 2004) recently, including watermarking, code obfuscation, integrity verification, wrappers and some hardware-based protection technologies. It is important to keep in mind that none of the above approaches can guarantee protection against attacks. A combination of the different protection techniques such that each masks the weaknesses of the others can provide better protection. Software-based protection techniques are explained below.

11.5.1 Encryption Wrappers

In encryption wrapper systems, software is encrypted and has to be decrypted before use. To be efficient, only critical parts of a program are encrypted. These parts are decrypted at the run-time before use. In other words, an adversary has to run the program in order to get the decrypted image of the program.

To prevent the attacker from getting a snapshot of the whole unencrypted software, only the codes that will execute in the system should be decrypted, and the other parts of the software should stay encrypted. This cannot guarantee protection since an adversary can still take many system snapshots, compare them and bring together the total unencrypted program.

In addition, decryption keys have to be protected from disclosure. The main attack methods of an adversary are run-time tools, such as debuggers or memory dumps, or running and analyzing the program in a virtual machine environment. To prevent attacks, the use of run-time tools has to be limited. At the same time, various defensive mechanisms can be used to make it more difficult to run and analyze a program in a virtual machine environment. Encryption wrappers can also use compression together with encryption to reduce the storage usage and to increase the security of the system.

An encryption wrapper is an effective protection solution. Attackers need to design more complex attack tools and consume more time to analyze and obtain an image of the decrypted program. However, encryption wrappers also add overhead for decryption in run time.

11.5.2 Code Obfuscation

A program can be decompiled into source code with the help of reverse engineering tools. Code obfuscation is a technique that can prevent reverse engineering and tampering attacks by transforming the original program into some code that does not have a consistent flow. The obfuscating transformations should maintain the program's functionality and have only moderate impact on the code performance and memory usage. At the same time, obfuscating transformations must be resilient to various attacks from automatic reverse engineering tools. The quality of obfuscating transformations may be classified and evaluated with respect to their *potency* (to what degree is a human reader confused?), *resilience* (how well are automatic deobfuscation attacks resisted?) and *cost* (how much overhead is added to the application?).

There are several kinds of obfuscation transformation (Collberg and Thomborson, 2002): *layout transformation, data transformation, control transformation* and *preventive transformation*. Layout transformation modifies the physical appearance of the code by removing the format of the code (e.g. nested conditional statements) and replacing the names of the important variables with random strings. Once the original formatting is lost, it is difficult to recover. This is only effective in combination with the other transformations. Data transformation changes the data structures used in the program, which includes changes to the way data are stored in the memory, how stored data are interpreted, how data are grouped and how data are ordered. Control obfuscation aims to manipulate the control flow of the program, e.g. changing how statements are grouped together, the order in which statements are executed and reorganizing the loop or block structures in a program.

Preventive transformations aim to make it more difficult for the deobfuscators to get the code of the program, making automatic deobfuscation techniques more difficult and exploiting known weaknesses of deobfuscators.

Obfuscation can be done at source-code and assembly levels. The transformations at the source-code level are more widely used, but if the source-code level transformations are not well designed, they can be undone by the code optimizer in the compiler and an adversary can easily find where the transformations were applied. Obfuscation at the assembly level works better since it can effectively hide the binary operation.

To evaluate the quality of the obfuscation transformations, potency and resilience (Collberg *et al.*, 1997) are two main factors to be considered. In other words, the quality is measured by the degree to which the transformed code is more obscure than the original, and by how well the transformed code can resist automated deobfuscation attacks. The measurement of resilience includes the programmer's effort to create a deobfuscator and the time and space required to run the deobfuscator. The level of security an obfuscator adds to an application depends on the sophistication of the transformations employed in obfuscation, the power of the available deobfuscation algorithms and the amount of resources available to the deobfuscator. In order to get the best result in terms of obfuscation, several of the obfuscation transformation techniques mentioned above can be combined to work together.

11.5.3 Software Watermarking and Fingerprinting

Software watermarking techniques have great implications for protecting intellectual copyright by discouraging piracy and illegal copying of digital items. Software watermarking is a special data structure, embedded into the software, which is used to show proof of ownership/authorship of the software. Fingerprinting is a special type of watermarking for identifying traitors who distribute the software illegally.

The behavior of the watermarked program should be affected if the watermark is distorted or destroyed. For protection, a watermark should be hidden using, for example, steganography. The watermark should be robust (hard to remove). In other words, a watermark must be able to withstand various attacks so that proof of ownership or the origin can still be extracted even after heavy attack. However, in order to provide tamper evidence, a fragile watermark may be desirable, where even a small alteration will destroy the watermark.

Fingerprinting is a kind of watermarking that embeds a unique message in each instance of the software for traitor tracing. In other words, it is distributed with different licensing messages. The disadvantage (Atallah *et al.*, 2004) is that if the adversary can gain access to several fingerprinted copies of an object, the location of the fingerprints can be determined by comparing them, and then the original program may be reconstructed by bypassing the fingerprinting.

Software watermarking includes static watermarking and dynamic watermarking. In static watermarking, the marks are stored directly in the data or code sections and can be read without running the program. In dynamic watermarking, the marks are stored in the run-time structures of the program and can be read only when the program is running. For both static and dynamic watermarking, a secret key is needed to read the marks.

Watermarks can also be used to enable the tracking of software copies and to demonstrate authorized possession or the fact that software has been pirated. But if the adversary can obtain the secret key, he can access information about the copyright owners and users that are licensed to use the software, which is stored in the watermarks. Hence, it should be noted that watermarks do not address reverse engineering or authorized execution issues (Atallah *et al.*, 2004).

11.5.4 Guarding

Using only a single protection scheme adds vulnerability because a single protection point, no matter how effective it is, can be located and compromised. Instead, to provide robust protection, multiple (possibly simple) protection techniques should cooperate together to resist tampering of software and to enforce security policies. These small security units are called *guards*.

A guard (Chang and Atallah, 2001) is a piece of code responsible for performing certain security-related actions during program execution. Guards are inserted into the software code to protect a specified region of code. Cross-guarding can be implemented between code sections in the software and other objects (e.g. hardware). Guards are triggered to action by unauthorized tampering of the protected program. Otherwise, a guard must not interfere with the program's basic functionality.

The guards are programmed to provide protection through, for instance, program code checksum, code repairing and so on. A large number of guards can form a network (Chang and Atallah, 2001) to reinforce the security of each other by mutual protection. Software-based guards can also be used together with hardware-based protection to ensure that the protected software can only be executed in an authorized environment.

Guards can provide multiple layers of defense, for instance with self-healing capabilities, diversity and random execution, to eliminate attacks. Guards should be resilient. To provide resistance to different attacks, a slight change of the script can yield a significantly different structure and code flow. Using high-level scripts, developers can select which specific guard instances to insert, and which specific sequence to use for transformations. Guards allow the developer to have precise control over the placement of protection code. Guards' reactions to attack can be flexible if needed. In other words, when attacks are detected, responses from the guards depend on the software publisher's business model and the expected adversary.

11.6 Tamper Resilience: Hardware Protection

Nodes in the sensor network are susceptible to physical attacks because they are deployed in an accessible environment and in a distributed manner. One of the physical attacks on wireless sensor networks is node capture, where an adversary can gain control of some nodes through direct physical access. Recently, with the development of hardware-tampering technology, sensor nodes may be compromised by a well-trained attacker in a very short time. After a node is compromised, secret information in the node may be disclosed and the behavior of the node is controlled by unauthorized access. All these may pose a great threat to the security of the whole network. To resist an attack, both software and hardware protection can be applied. Some approaches use hardware to make it difficult for an adversary to access the information stored in the sensor nodes, others may use both hardware and software to protect the nodes. With rapid decreases in hardware costs, hardware-based protection will become more effective for wireless sensor networks.

To protect the node from physical attack, the node can choose to destroy all the data and keys by itself when it senses a possible attack. The network can efficiently avoid potential attack by killing the compromised node itself. All the node needs to do is to detect a possible intrusion. As soon as an attack is sensed by using a low-frequency sensor, the information in the sensor can be made invisible to the attacker or the circuitry can be destroyed to prevent the attacker from accessing the information in the node.

There are many kinds of physical attack directed at wireless sensor networks, including getting information from the sensor by manual probing, laser cutting and power analysis. For example, some tamper resistance devices such as smartcards (Komerling and Kuhn, 1999) can be reverse engineered using chip-testing equipment. Helped by special techniques (Anderson and Kuhn, 1997), such as differential fault analysis, chip rewriting and memory remanence, the cost of reverse engineering can be reduced heavily. To protect the security of networks, methods that can make such attacks difficult have been introduced (Komerling and Kuhn, 1999). For example, we can use randomized multithreading to introduce more nondeterminism into the execution of algorithms; inserting random time delays between any observable reaction and critical operations can increase the difficulty of attack; adding additional metallization layers that form a sensor mesh above the actual circuit and that do not carry any critical signals remains one of the most effective irritants to microprobing attackers.

We can categorize hardware-based protection schemes that use these approaches into two classes (Atallah *et al.*, 2004): using *tamper-resistant processors* for protection and using lightweight hardware such as *tokens*. In the tamper-resistant processor approach, the systems rely on secure processors to prevent illegal software duplication, unauthorized software modification and unauthorized software reverse engineering. Such secure processor architectures adopt some features to enable a secure environment where only authorized and untampered hardware and software can exist. Security is achieved by protecting data located in the hardware devices by means of encryption and dynamic integrity checking when the data and instructions are brought into the tamper-resistant processor. When executing applications, the architecture ensures that sensitive data and instructions of the applications will not be disclosed at any time and the software integrity is always guaranteed. Secure processor architectures can be used not only to prevent possible software-based attacks on protected applications but also to prevent many hardware attacks.

Another hardware-based protection mechanism is to use more lightweight hardware such as hardware tokens or tamper-resistant smartcards to prevent various attacks. In this kind of protection mechanism a user license is embodied by a copy-resistant piece of hardware. The software or the operating system checks for the presence of the token and refuses to run without it. Mechanisms for checking the hardware token are based on a cryptographic key that is never stored or used outside the tamper-resistant token.

Hardware-based protection has both advantages and disadvantages. The most important advantage of hardware-based protection is that the application is running on a trusted processor, which can prohibit access to its memory and provide greater protection against attacks. The software is stored in the system storage in encrypted form and it can only be decrypted by the processor internally before execution, which prevents any user having full control of the system from examining the clear text instruction. More importantly, the data communicated between the processor and the memory are all encrypted to prohibit reverse engineering of the code. Another advantage of hardware-based protection is that only a program that has passed the encryption and dynamic integrity checking can be brought into the tamper-resistant processor.

However, there are many disadvantages to hardware-based protection. Compared with software-based protection, hardware is more difficult to modify and update and more time is needed to install, integrate and test hardware. Another drawback of hardware-based protection is the cost of hardware, which is much higher than using software protection. In addition, since every time the data are brought into the tamper-resistant processor, encryption and dynamic integrity checking are needed, hardware-based protection is more costly with respect to energy, delays and system resources like memory and processing time.

The bottom line is that both hardware- and software-based anti-tamper techniques are used to prevent the unauthorized modification and use of software, and both have their advantages and disadvantages. Software-based techniques include watermarking, code obfuscation, integrity verification and wrappers. The main drawback of the software anti-tamper approach is that the application is running on a host that cannot be trusted (Atallah *et al.*, 2004). But since hardware-based anti-tamper techniques can use a trusted device to guarantee the safety of the program, by using a combination of both software- and hardware-based anti-tamper solutions, we can achieve stronger protection of the nodes.

11.7 Availability and Plausibility

Availability means the ratio of the duration that a network is functioning properly during a given time period and the time period. For a wireless sensor network, the availability implies the percentage of sensors that can get the information and send it to the sink without failure.

In a secure network availability should be high. All the security techniques we explain in this book can increase the availability of ad hoc and sensor networks.

In wireless ad hoc sensor networks, the verification of data consistency may be required for the received and aggregated data at the sink node. This requirement is called *data plausibility*. Checking the plausibility of the received data is a useful method for defending against compromised nodes, and authenticity, integrity and confidentiality may not guarantee plausibility. The reason is that a message may contain false data while the sender may be legitimate. For example, an intruder may trick a light sensor by covering it with dark clothes. A defense mechanism based on redundancy can be developed against these kinds of tricks. The legitimacy of messages should be ensured by their consistency with messages coming from the neighboring nodes. Therefore, the plausibility of the received data can be checked by comparing the data received from all the neighbors with the state of the surrounding environment obtained by sensors or by predefined rules.

11.8 Review Questions

11.1 Outline the methods used to support privacy in wireless sensor networks. Give the reasons for using each method.

11.2 Outline the detection techniques used in the intrusion-detection system and the intrusion-detection system architecture in wireless ad hoc networks.

11.3 What are the characteristics of intrusion-detection systems in wireless sensor networks?

11.4 What is the relationship between a password mechanism and network security?

11.5 Outline the user characteristics related to password design.

11.6 What is a knowledge-based authentication system and how does it work?

11.7 How many software-based protection mechanisms are used for software protection? How do these mechanisms work?

11.8 Compare the advantages and disadvantages of software-based and hardware-based protection mechanisms.

11.9 What are data availability and plausibility? How can we achieve them?

12

Secure Routing

Security threats against routing schemes are explained in Chapter 8. In this chapter we introduce the counter measures against those threats. First we summarize the approaches for secure routing. In this summary the focus is on wormhole attacks, sybil attacks, selective forwarding and secure multicasting/broadcasting. Routing techniques that improve the performance of security measures for other services are also elaborated. Selected secure routing protocols are explained in the second part of this chapter.

12.1 Defense Against Security Attacks in Ad Hoc Routing

Since routing is one of the important challenges in ad hoc and sensor networks, it has attracted many researchers and been studied extensively. Most of the research has focused primarily on quality of service guarantees for ad hoc networks and energy efficiency for sensor networks. However, many ad hoc and sensor network applications are designed to be deployed in hostile environments and are subject to hostility. Therefore, security should also be considered one of the primary factors influencing the design of routing protocols.

There are three approaches to designing a secure routing protocol (Parno *et al.*, 2006):

- attack prevention;
- attack detection and recovery from the attack;
- resilience to security attacks.

Routing protocols can be designed such that an adversary cannot compromise nodes/messages or make the routing scheme dysfunction. This is the most effective approach with respect to the cost of the security scheme and effectiveness in defense against the threats. Therefore, most of the techniques fall into this category. Preventive approaches are designed to counter known threats and may not be effective against new threats. Detection schemes for misbehaving or malfunctioning nodes can be designed in a more generic fashion. On the other hand, they can be more costly than preventive approaches. Finally, routing can be designed such that it still delivers the data packets to the destination when there is an attack. Such resilient techniques are also costly. We give examples of both detection and resilience techniques.

Security in Wireless Ad Hoc and Sensor Networks   Erdal Çayırcı and Chunming Rong
© 2009 John Wiley & Sons, Ltd

12.1.1 Techniques Against Wormhole Attacks

Wormholes are difficult to detect because an adversary passes the packets to a distant point from the point at which they are received by using a single hop out-of-band channel. This channel cannot be listened to by the network. Moreover, the real copy of the packet reaches the point that receives the replayed copy later than the replayed copy. Therefore, the replayed copy is fresher than the real copy.

Detection mechanisms against wormhole attacks can be based on temporal and spatial analysis of the packets. Geographical and temporal packet leashes introduced in Hu *et al.* (2003) follow this approach. A *geographical leash* scheme assumes that nodes are loosely synchronized and location aware. The source node S includes its location l_S and the packet transmission time t_S as the geographical leash into its packet P_S sent to destination D.

$$S \rightarrow D : l_S, t_S, P_S$$

The clocks of the nodes in the network are synchronized to within $\pm\Delta$. The upper bound for the distance between two nodes is d_b, and is based on the transmission range of the nodes. The node localization error upper bound is also given as δ. Similarly, an upper bound for the velocity in transmitting signals v is also known. Then every node i that forwards the packet, which is at location l_i, and receives the packet at time t_i can check the following condition.

$$d_b \leq |l_i - l_S| + 2v \times (t_i - t_S + \Delta) + \delta \qquad (12.1)$$

If this condition does not hold, it indicates that the packet has been received by the node i earlier than expected and the network may be subject to a wormhole attack. When the average distance between nodes is not long enough, and the number of hops in the normal path is not high enough, this technique may not detect wormholes.

Temporal leashes use only the transmission and reception times of the packets for detecting wormholes. When a node A sends or forwards a packet to another node B, it also includes the transmission time t_A into the packet P_A.

$$A \rightarrow B : t_A, P_A$$

Node B checks the difference d_{AB} between the transmission time t_A and reception time t_B of the packet. If d_{AB} is shorter than a given threshold θ, it may indicate a wormhole attack. Temporal leashes require tight time synchronization.

When directional antennae are available in nodes, they may also be used in the detection of wormholes (Hu and Evans, 2003), as shown in Figure 12.1, where the data packets transmitted by node a are received by a malicious node $w1$, conveyed to another malicious node $w2$ through a wormhole and replayed by node $w2$. The replayed packets are received by node f. Therefore nodes a and f believe that they are neighbors.

Let's assume that both nodes a and f are equipped with a directional antenna that has six sectors, and the sectors of both nodes are aligned, i.e. Sector 1 of node a is in the same direction as Sector 1 of node f. Therefore, the packets transmitted at Sector 4 of node a should be received at Sector 1 of node f. However, the packets replayed by the wormhole are received at Sector 2 of node f. This indicates a wormhole attack and can be detected by node f if the packets also include data about the sector from which they were transmitted.

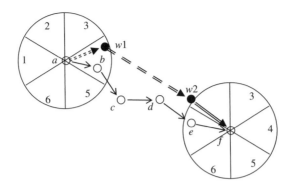

Figure 12.1 Directional antennae to detect wormhole attacks

Attackers can also adapt to directional antennae by replaying the packets in the same sector as that in which they are received. However, the capabilities of attackers are reduced even with this simplest form of the approach. More sophisticated schemes based on the same idea are introduced in Hu and Evans (2003).

12.1.2 Techniques Against Sybil Attacks

To defend against sybil attacks, the identities of every node should be verified. This can be done either directly or indirectly. In direct validation a node directly verifies whether the identity of a neighboring node is valid. For example, a node may assign each of its neighbors a separate channel to communicate and ask them to transmit during a period. Then it checks these channels in a random order within that period. If a node is transmitting in its assigned channel, the node is a physical node. If no transmission is detected on a channel, it indicates that the node assigned to that channel may not be a physical node (Newsome *et al.*, 2004).

In indirect validation another trusted node provides the verification for the identity of the node. For example, every node may share a unique key with the base station. When two nodes need to establish a link between them, they verify each other's identity through the base station by using these keys (Karlof and Wagner, 2003). At the same time they can be assigned a session key. Nodes can also be allowed to establish links with a limited number of neighboring nodes. Thus, compromised nodes can only communicate with a limited number of verified neighboring nodes, which also limits the impact of sybil attacks.

Random keys assigned to nodes also provide security against sybil attacks. Since a limited number of keys is available to each node, nodes do not have enough keys to generate multiple identities (Newsome *et al.*, 2004).

12.1.3 Techniques Against Selective Forwarding

Preventing wormholes, sink holes and sybil attacks cannot guarantee to mitigate black hole and selective forwarding attacks. A compromised node can still act as a black hole or drop selected packets. There are two approaches to defending against selective forwarding: detecting the nodes that selectively forward, and developing routing schemes that are more resilient and can deliver packets even when there is a selective forwarding attack.

One approach to detecting the nodes that selectively forward is based on acknowledgements (Yu and Xiao, 2006). Every intermediate node that forwards a packet waits for an acknowledgement from the next hop. If the next hop node does not return the same number of acknowledgements as the number of packets sent, the node generates an alarm about the next hop node. However, compromised nodes can also generate acknowledgements for the packets that they dropped, which make this scheme fail. Moreover, a malicious node can generate fake alarms to organize a DoS attack. Authentication schemes and encryption can be used to prevent these kinds of malicious behavior (Yu and Xiao, 2006). Link layer acknowledgements can also be complemented by end-to-end reliability schemes.

Multipath routing can be an effective way to mitigate selective forwarding and black hole attacks (Karlof and Wagner, 2003). This requires at least link-disjoint paths, where two paths may share some nodes but no link. Of course, node-disjoint paths, where two paths do not have any node in common, are better and reduce the risk of selective forwarding attack compared to link-disjoint paths. However, disjoint paths are not always available, and when paths are not disjoint, if the selectively forwarding node is the node common to all the paths, then the attack can become as effective as in single-path routing.

12.1.4 Secure Routing in Sensor Networks

The traffic regime in wireless sensor networks is mainly either broadcasting or multicasting. In a typical scenario the base station broadcasts administrative messages and queries in the network, and multiple sensors reply to the queries broadcast by the base station. Therefore, the traffic from the base station to the nodes, i.e. the downstream traffic, is either broadcast or 1-to-M, and the traffic from the nodes to the base station, i.e. the upstream traffic, is M-to-1. When actuators or multiple base stations are used, the upstream traffic can be characterized as M-to-N. The security issues for the routing schemes in such traffic regimes are different from the issues for 1-to-1 routing. Therefore, security schemes specifically designed for the following classes of routing are required for wireless sensor networks:

- secure broadcasting for the downstream traffic;
- secure multicasting for the downstream traffic;
- secure data aggregation when routing from multiple nodes to a base station;
- secure data aggregation and multicasting when routing from multiple nodes to multiple base stations or actuators.

Since these relations are typical in sensor and actuator networks, we explain them under the topic of 'secure routing in sensor networks.' Secure broadcasting and multicasting are also required for conventional ad hoc networks, and the approaches explained in this section are also applicable to them.

12.1.4.1 Secure Multicasting and Broadcasting

In secure multicasting and broadcasting one of the major challenges is the group key and trust management, which can be achieved in a centralized, clustered or distributed manner. The group key and trust management are explained in Section 10.3. In central group key management a central key manager manages the keys for the entire network, including the node that broadcasts or multicasts. Logical key trees (Di Pietro *et al.*, 2003) are an example of a central group key management scheme. In the clustered approach, the key management responsibility is shared among multiple key managers, where each manager holds this role for a group of nodes. In the distributed approach, group key management is achieved by the node that multicast. TESLA and μTESLA are examples of this approach. μTESLA is explained in Section 13.1.2.

12.1.4.2 Secure Data Aggregation

Every sensor node is a separate entity that sends its messages to the base station. Therefore, the networking regime can be perceived as 1-to-1 between one sensor node and the base station. However, the relationship between sensor nodes and the base station is M-to-1 because the data traffic in a sensor network is both temporally and spatially correlated due to the overlapping sensing regions, and events that cover large areas can be detected by multiple sensors. Multiple sensors may report the same event or reply to the same query almost at the same time. In addition to this, it is more efficient to aggregate multiple packets into a single packet while they are being conveyed in a network. The nodes where the packets are aggregated, i.e. aggregators, can be any node and can be selected previously. Aggregation creates a reverse multicast tree that ends at the base station.

Data aggregation increases the vulnerability of a sensor network for the following reasons:

- compromising an aggregator enables the forging of data coming from all the nodes before the aggregator;
- the integrity of a single data packet represents the integrity of multiple data packets aggregated into it.

Therefore, the aggregators are primary targets in organizing wormhole, sink hole, black hole, selective forwarding and sybil attacks. When these attacks are accomplished on aggregators, their impact becomes higher. All the techniques designed to counter these types of attack also improve the security for the data aggregation scheme.

Techniques that prevent adversaries from analyzing the traffic can hide the aggregators. When the aggregators are covert, attacks that target them cannot be organized easily. In addition to this, authentication and encryption schemes provide additional measures to protect the aggregators and aggregated messages from being compromised.

Finally, the aggregators can check the consistency of every data packet before aggregating them. To do this, a statistical analysis can be carried out. Similar approaches have also been introduced for the security of other services such as secure event boundary detection and secure localization, explained in the following chapter.

12.1.5 Routing Schemes That Enhance Security

Security is not envisaged only to secure the routing scheme. Routing schemes are sometimes designed such that they can contribute to the security measures for anonymity, privacy and attacks involving traffic analysis. For example, a random-walk-based routing scheme may be designed to thwart traffic analysis.

In a *random walk* strategy, nodes do not always forward the packets to their next hop node. Instead they may occasionally send some of their packets to a randomly selected node (Deng *et al.*, 2004). This helps to mitigate traffic analysis attacks trying to detect base stations and aggregators by monitoring the traffic rates at the links. In an alternative strategy, when a node forwards a packet towards the base station, its neighbor also forwards a fake packet to a randomly selected node (Deng *et al.*, 2004). The fake packets are dropped after a certain number of hops. This approach prevents adversaries from detecting the base stations and aggregators by examining the timings of data transmissions.

Another random-walk-based technique is the *greedy random walk* (Xi *et al.*, 2006), where both the base station and the source node start a random walk. When the routes created by these random walks intersect, the packets sent by the source node follow the route created by the random walk started at the base station.

Flooding may also be considered a routing technique resilient to traffic analysis (Walters *et al.*, 2006). In Ozturk *et al.* (2004), flooding-based techniques that provide additional privacy measures compared to single-path routing are introduced:

- **Baseline flooding:** this is the flooding scheme that we introduced in Section 5.2.1. Every node broadcasts every packet that it receives, and does this only once to eliminate loops.
- **Probabilistic flooding:** a subset of the nodes is not involved in flooding and drops the packets that they receive. This technique cannot guarantee the delivery of every packet.
- **Flooding with fake messages:** an adversary can still analyze the traffic and monitor individual packets in baseline and probabilistic flooding techniques. To mitigate this risk, random nodes generate fake messages and flood them together with the real packet in this technique.
- **Phantom flooding:** this technique has two phases. In the first phase, a packet makes a given number of hops by using one of the random or direct walk techniques. In the second phase, at the end of the walk phase the packet is flooded from the node that it reaches at the end of the walk phase.

When these techniques are supported by other schemes for anonymity, such as decentralizing sensitive data, securing communication channels and using mobile nodes, traffic analysis becomes very difficult for adversaries (Walters *et al.*, 2006). Techniques for anonymity have already been explained in Chapter 11.

12.2 Secure Ad Hoc Routing Protocols

In this section we explain the secure ad hoc and sensor network routing protocols listed in Table 12.1. Some of these protocols are also analyzed in Fonseca and Festag (2006). The list of protocols in this chapter is not exhaustive. There are many others in the literature. Our aim is to provide a better insight into the requirements for a secure ad hoc routing protocol. Therefore, we have selected a list of protocols that are representative of various approaches in the literature.

Table 12.1   Secure ad hoc routing protocols

Name	Routing scheme	Security services
Intrusion-tolerant routing in WSNs (INSENS)	– Routing scheme for fixed sensor networks – Multipath link state routing – Base station computes the routes and broadcasts them	– Resilience to physical attacks and compromised nodes – Authentication – Resilience to DOS-style flooding attacks – Symmetric cryptography
Authenticated routing for ad hoc networking (ARAN)	– Directed diffusion like routing protocol where gradients are set during flooding-based route request phase – Designed for ad hoc networks	– Certificate management – Data integrity – Asymmetric cryptography
On demand secure ad hoc routing (ARIADNE)	– Dynamic source routing – Designed for ad hoc networks	–Authentication – Resilience to node compromises – Resilience to active-1-x and active-y-x attackers (see Section 8.2) – Symmetric cryptography
Watchdog Pathrater	– Dynamic source routing – Designed for ad hoc networks	– Detection and recovery scheme
Secure ad hoc on-demand distance vector (SAODV)	– Ad hoc on-demand distance vector routing – Designed for ad hoc networks	– Import authorization – Source authentication – Integrity – Data authentication – Asymmetric cryptography and hash chains
Secure link state routing protocol (SLSP)	– Link state routing – Designed for ad hoc networks	

12.2.1 Intrusion-Tolerant Routing in Wireless Sensor Networks (INSENS)

INSENS (Deng *et al.*, 2003) adheres to three design principles. First, only the base station can broadcast or multicast. Individual nodes are only allowed to unicast to the base station. The base station constructs the forwarding tables for every node. The entire network sends its neighborhood information to the base station and the base station creates the forwarding tables for the nodes and distributes them to the nodes. Forwarding tables are constructed such that routes are multiple path routes for each source. Second, all control and routing information must be authenticated. Finally, symmetric keys are used for authentication. This comprises route discovery and data forwarding phases. The route discovery phase is also subdivided into three steps: route request, route feedback and route computation and dissemination.

12.2.1.1 Route Discovery Phase

Whenever the base station needs to construct the forwarding tables it broadcasts a *route request* message. The route request message is flooded into the network. Every node that receives a route request message first appends its identity to the message and retransmits the message. At the same time, it records the identity of the sender in its neighbor set. When a node receives another copy of the same route request, it only records the sender and drops the request.

To prevent spoofing in this process, a broadcast authentication scheme similar to TESLA is used. A set of numbers $K = \{K_0, K_1, \ldots, K_n\}$ is generated, such that $K_i = F(K_{i+1})$, where F is a one-way cryptographic hash function, and $0 \le i < n$. The entire network knows F and K_0 at the time of deployment. Only the base station has all the numbers in K, and it uses K_i to broadcast the ith route request. The nodes that receive the ith route request generate K_{i-1} from the K_i available in the message by using F, and compare it with the K_{i-1} known to them before receiving the message. If they match, the message is authentic. Since F is a one-way function no other nodes can generate K_{i-1} by using K_i. Therefore, a compromised node cannot spoof the base station by generating new keys. However, it can replay the route request message, which can damage the process only in the downstream from the compromised node.

Every node is also configured with a separate secret key shared only with the base station. The nodes generate a MAC on the complete path after adding their identity to the path and append their MAC to the message before transmitting it. The base station later uses these MACs to verify the nodes in the path. If a node is compromised, only one secret key is revealed, so an attacker cannot compromise the entire network or organize sybil attacks by compromising a single node.

After a node transmits an incoming route request message, it waits for a time interval t. During t, it receives copies of the same route request from its neighbors and records both the identities and MACs of the neighbors in its neighbor set. At the end of t, it sends a *route feedback* message to the base station through the reverse path to that on which it received the route request message. The route feedback message includes both the neighbor set and another MAC generated on the neighbor set by using the key of the node. The latter MAC ensures the integrity of the neighbor set information.

The base station populates the topology data from incoming route feedback messages. Since every node sends its neighbor set, the integrity of the reported neighbor sets can be compared to each other and inconsistencies can be detected. The nodes are also verified based on their MACs. When the topology of the network is constructed and verified, multipath routes and forwarding tables for every node are computed by the base station. Then the forwarding tables are disseminated, starting from the first-hop nodes to the base station.

12.2.1.2 Data Forwarding Phase

The forwarding table of a node is comprised of several entries, i.e. one record for each route that the node is a part of. Each record has three tuples:

<destination, source, immediate sender>

Destination is the node that the message is sent to. Source is the node that creates the message. Immediate sender is the node that forwards this message to the owner of the forwarding table. For example, let's assume that one of the paths in the multipath routing tables is as follows:

$$S \text{ to } D : S \rightarrow a \rightarrow b \rightarrow c \rightarrow D$$

For this path the forwarding table of a includes a record $<D, S, S>$, the table of b has $<D, S, a>$ and the table of c has $<D, S, b>$. When a node receives a packet, first it checks its forwarding table. If it finds a tuple that matches the destination, source and the immediate sender in the packet, this indicates that the node is on the route and should forward the packet to the next hop. Therefore, it just replaces the immediate sender field in the packet with its own id and broadcasts the message.

12.2.2 Authenticated Routing for Ad Hoc Networking (ARAN)

ARAN (Sanzgiri *et al.*, 2002) was designed for ad hoc networks rather than wireless sensor networks. Although it is scalable and adaptive enough for many sensor network applications, it is based on asymmetric encryption and therefore does not fit well the hardware constraints of typical sensor nodes. Another important feature that makes it different from the secure sensor network routing protocols is that ARAN requires a trusted certificate server that both authenticates the nodes and assigns temporary certificates to them. The public key of the trusted certificate server is known by every node. Apart from this, ARAN assumes that there is a mechanism to authenticate nodes when they apply for the certificates. That mechanism is not specified by the protocol.

ARAN provides mechanisms for certification, integrity and nonrepudiation. When a node A accesses the network for the first time or needs a certificate for route discovery, it requests the certificate from the trusted server T. The server T first authenticates the node A and sends a certificate to it:

$$T \rightarrow A : certificate_A$$

where $certificate_A = \{IP_A, K_{A+}, t, e\}_{<K_{T-}>}$

IP_A is the IP address of node A,
K_{A+} is the public key of A,
t is the time that the certificate is created,
e is the time that the certificate expires,
K_{T-} is the private key of T.

A node S that has a valid certificate can start a route discovery for another node D by broadcasting a route discovery packet (RDP):

$$S \rightarrow \text{broadcast} : \{RDP, IP_D, certificate_S, N_S, t\}_{<K_{S-}>}$$

where N_S is a nonce, which is the sequence number, i.e. the source node S monotonically increases the nonce each time it performs a route discovery to ensure the freshness of the reply message expected from the destination D.

When a node receives an RDP message, it first decrypts the message and then records the neighbor that sends the message as the next hop node for the source node of the message. If the node receives a reply message for this RDP, it just forwards the reply to the neighbor in this record. Finally, it encrypts the message by using its private key, appends its certificate and broadcasts the message.

$$B \rightarrow \text{broadcast} : \left\{ \{RDP, IP_D, certificate_S, N_S, t\}_{<K_{S-}>} \right\}_{<K_{B-}>} , certificate_B$$

Note that every intermediate node decrypts the received message, encrypts it again by using its own private key and appends its certificate to the message before forwarding. To decrypt the message a node needs the public key of the neighbor that it receives the message from. That public key is in the certificate appended to the message, which is encrypted by the private key of the trusted server. Every node knows the public key of the trusted server, and certificates are issued encrypted by the trusted server, as explained above. For example, if node C receives the RDP message forwarded by node B, the message that node C broadcasts looks like this:

$$C \rightarrow \text{broadcast} : \left\{ \{RDP, IP_D, certificate_S, N_S, t\}_{<K_{S-}>} \right\}_{<K_{C-}>} , certificate_C$$

Each node authenticates the previous node in the route because messages are signed at each hop. In addition to this, an RDP does not contain a hop count or record a source route. Therefore, malicious nodes do not have the opportunity to redirect traffic by tunneling or modifying route sequence numbers, hop counts or source routes.

When destination node D receives the route discovery message from the last node in the route, let it be C for our example, it first verifies the source's signature and then prepares a reply (REP) message and unicasts it to C:

$$D \rightarrow C : \{REP, IP_S, certificate_D, N_S, t\}_{<K_{D-}>}$$

Like RDP messages, REP messages are also decrypted and encrypted, i.e. signed by, all the intermediate nodes. Every intermediate node also replaces the certificate appended to the message. The main difference in forwarding the REP message to the source node S from RDP message delivery is that they are not broadcast but unicast by every intermediate node to the next hop node recorded while conveying the RDP message.

$$C \rightarrow B : \left\{ \{REP, IP_S, certificate_D, N_S, t\}_{<K_{D-}>} \right\}_{<K_{C-}>} , certificate_C$$

When the source node S receives the REP message, it verifies the destination's signature and the nonce. After this the route establishment is completed.

Every intermediate node maintains the route in its route table. When no traffic occurs on an existing route for longer than the route's lifetime, it is deactivated in the route table. When a message is received to be forwarded in a deactivated route, the node generates an error (ERR) message and forwards it to the next hop in the reverse path towards the source node. ERR messages are also sent when a node finds out that a link is broken for reasons like node movement. Every ERR message is signed and unicast. It has the format below when an intermediate node C forwards it to another node B:

$$C \rightarrow B : \{ERR, IP_S, IP_D, certificate_C, N_C, t\}_{<K_{C-}>}$$

This message is forwarded to the source node without being modified. When it reaches the source node, the route is also deactivated by the source node.

Like routes, certificates can also be revoked. To do this, the trusted certificate server broadcasts a message that announces the revocation of $certificate_R$.

$$T \rightarrow \text{broadcast} : \{REVOKE, certificate_R\}_{<K_{T_-}>}$$

Revocation messages are stored until the revoked certificate expires.

12.2.3 On-Demand Secure Ad Hoc Routing (ARIADNE)

ARIADNE is a secure dynamic source routing protocol (Hu *et al.*, 2005). It provides secure route discovery and maintenance services by using various approaches, i.e. TESLA, MAC and digital signatures. When ARIADNE route discovery is used with TESLA, each hop authenticates the information in the 'route request'. The target buffers the 'route request' and does not send back a 'route reply' until the intermediate nodes release their TESLA keys. Then it completes the authentication of both the source node and the intermediate nodes and sends back a 'route reply' on the reverse path to the one on which the 'route request' was conveyed.

ARIADNE assumes that every communicating source node S and destination node D share the MAC keys K_{SD} and K_{DS}. Moreover, every node has a TESLA one-way key chain and all nodes know the first key in the key chain of each other node.

The ARIADNE route discovery process starts with a 'route request' that has the following fields:

- **Route request:** the code word, which indicates that the packet is a route request packet
- **Source node**
- **Destination node**
- **Id:** the identification of the route request
- **Time interval:** a pessimistic estimation of the time required for delivery of the packet to the destination
- **Hash chain:** the hash value created by all the nodes in the route
- **Node list:** the list of nodes in the route
- **MAC list:** the list of the MAC values calculated by every node in the route.

The source node assigns the source, destination, id and time interval fields and they never change until the packet arrives at the destination node D. The hash chain, node list and MAC list change in every hop. The hash chain is computed first by the source node S as follows:

$$h_0 = MAC(K_{SD}, REQUEST|S|D|id|t_i) \tag{12.2}$$

After computing h_0, the source node initializes the node list and MAC list fields as empty lists and broadcasts the 'route request' message.

$$S \rightarrow \text{broadcast} : \{REQUEST, S, D, id, t_i, h_0, (\,), (\,)\}$$

Every node that receives the route request first checks the $<source, id>$ fields in its buffer. If this request has already been received, the new request is dropped. The node also checks the time interval. If it is too far in the future or the key associated with it is already disclosed, the packet is discarded. Otherwise, the receiving node modifies the hash chain h_i. Assume that A is a node one hop from the source node S. It computes h_1 as follows:

$$h_1 = H(A, h_0) \tag{12.3}$$

It also calculates its MAC value by using the next key K_{Ati} in the TESLA key chain, adds its address and the MAC value to the 'route request' message and broadcasts it:

$$M_A = MAC(K_{A_{ti}}, REQUEST|S|D|id|t_i|h_1|(A)|(\,))$$

$$A \rightarrow \text{broadcast} : \{REQUEST, S, D, id, t_i, h_1, (A), (M_A)\}$$

This process is repeated by every intermediate node. For example, the third intermediate node C calculates the hash chain and MAC and broadcasts the 'route request' message as follows:

$$h_3 = H(C, h_2) \tag{12.4}$$

$$M_C = MAC(K_{C_{ti}}, REQUEST|S|D|id|t_i|h_3|(A, B, C)|(M_A, M_B))$$

$$C \rightarrow \text{broadcast} : \{REQUEST, S, D, id, t_i, h_3, , (A, B, C), (M_A, M_B, M_C)\}$$

When the destination node receives the 'route request', it checks the validity of the request by determining that the keys of the time interval have not been disclosed yet, and the final hash chain is equal to

$$H(a_n, H(a_{n-1}, H(\ldots, H(a_1, MAC(K_{SD}, REQUEST|S|D|id|t_i)) \ldots)))$$

where a_n is the address of the node at position n and there are n nodes in the node list. If both of these conditions hold, the request is valid. Then the destination node D computes the destination MAC M_D, prepares a 'route reply' message and returns it along the source route obtained by reversing the sequence of hops in the node list of the 'route request' message.

$$M_D = MAC(K_{DS}, REPLY|D|S|t_i|(A, B, C)|(M_A, M_B, M_C))$$

$$D \rightarrow C : \{REPLY, D, S, t_i, , (A, B, C), (M_A, M_B, M_C), M_D, (\,)\}$$

In the reverse path, every node waits until it can disclose its TESLA key. After that it appends its TESLA key and forwards to the next hop in the reverse path.

$$A \rightarrow S : \{REPLY, D, S, t_i, (A, B, C), (M_A, M_B, M_C), M_D, (K_{C_{ti}}, K_{B_{ti}}, K_{A_{ti}})\}$$

When the source receives the 'route reply' message, it verifies that each key and each MAC are valid. If they are, it accepts the 'route reply' message. Otherwise it discards the message. After this, the route is maintained in the 'route cache' until a 'route error' message is received.

When an intermediate node B that tries to forward a message to the next node C in the route fails, this generates the following 'route error' message which is sent to the source node S along the reverse path.

$$B \rightarrow A : \{ERROR, B, C, t_i, M_B, K_{B_{ti-1}}\}$$

M_B is set to the MAC of the previous fields in the error message by using the next TESLA key K_{Bti} which will be disclosed at the end of t_i. Every node in the reverse path waits until the node that detects the error discloses its next TESLA key at the end of t_i. If the disclosed key verifies the MAC, the route is removed from the 'route caches'.

12.2.4 Watchdog Pathrater

The watchdog pathrater scheme (Marti *et al.*, 2000) is another secure routing scheme designed as an extension to DSR. It is not a preventive technique but is based on the detection and recovery approach. In this scheme every node runs two additional processes in the background, namely watchdog and pathrater. Watchdog maintains a buffer of transmitted packets. It also listens to the next hop nodes to observe if they forward the messages given to them, i.e. passive acknowledgement. When the next hop node forwards a packet, watchdog deletes it from the buffer. If a packet stays in the buffer for longer than a given time interval, watchdog increments a failure counter assigned to the node responsible for forwarding the packet. If the counter of a node becomes higher than a threshold, the node is marked as misbehaving.

Pathrater rates the links based on their reliability and knowledge of misbehaving nodes. Every node rates every other node in the network. When a link is used successfully, its rate increases. If a link break occurs, the rate of the link decreases. High negative numbers are assigned to nodes suspected of misbehaving.

Paths are rated by averaging the link ratings along the path. When the source node has multiple options to a destination, it selects the path with the highest path rate. Paths that contain misbehaving nodes are avoided. When there is no misbehaving-link-free path to the destination, the source node initiates a 'route request' process.

12.2.5 Secure Ad Hoc On-Demand Distance Vector (SAODV)

The secure ad hoc on-demand distance vector (SAODV) algorithm provides import authorization, source authentication, integrity and data authentication services for the ad hoc on-demand distance vector (AODV) algorithm. SAODV assumes that there is a key management system to assign public keys to the nodes and to verify the association between the node identities and public keys (Zapata and Asokan, 2002). Two mechanisms are used for the security services: hash chains to secure the hop count and digital signatures to authenticate the fields in the messages.

To secure the integrity of the hop count, a hash chain is formed by applying a one-way hash function H to a randomly selected seed value s. Before transmitting a route request (RREQ) or route reply (RREP) message, the source sets hash value h to seed s. The maximum hop count is assigned the time to live value *ttl*, and then top hash value *T* is computed by applying the hash function *ttl* times to seed s.

$$h = s \tag{12.5}$$

$$T = H^{ttl}(s) \tag{12.6}$$

When a node *i* receives a message after *i* hops from the source node, it first checks if the following condition holds:

$$T = H^{ttl-i}(h) \tag{12.7}$$

Since every intermediate node applies hash function *H* once to the hash value *h* in the message before relaying it, when *H* is applied *ttl − i* times to the current *h*, it should give the top hash value *T*. Otherwise it indicates that either the hash value *h* or the hop count *i* is incorrect. After this check, node *i* applies *H* to *h* and forwards it.

$$h = H(h) \tag{12.8}$$

To protect the integrity of the other fields in the message the source node signs everything but the hop count and hash value *h* fields, which are modified by every intermediate node. This would be enough if only the destination nodes were allowed to return an RREP for an RREQ message. However, if an intermediate node has a fresh route to the destination, it can also send back an RREP without further forwarding the RREQ. To do this, the intermediate node should be able to sign the RREP on behalf of the destination.

The first solution does not allow the intermediate nodes to generate RREQ messages. This approach is not the optimal one but is secure. In the second solution the node that generates RREQ messages includes the digital signature that can be used by the intermediate nodes as well as an RREP flag and prefix size. When an intermediate node generates an RREP message, the lifetime of the route is changed from the original one that the intermediate node received before. The intermediate node includes both the old lifetime that it received from the destination and the new lifetime in its RREP message. The old lifetime verifies the signature of the destination. The new lifetime is signed by the intermediate node.

12.2.6 Secure Link State Routing Protocol (SLSP)

The secure link state routing protocol (SLSP) (Papadimitratos and Haas, 2003) secures the link state discovery, i.e. neighbor discovery, and link state distribution, i.e. link state updates (LSUs), in the link state routing protocol. It does not attempt to secure the network from later stage attacks, such as, the integrity of the delivered data.

In SLSP each node is equipped with a public key E_v and a private key D_v, as well as with a one-way hash function *H*. Nodes are identified by IP addresses and they disseminate their LSU packets and maintain the topological information for the subset of nodes within *R* hops, which is termed a *zone*. Nodes periodically broadcast their certified key within their zone, so that the receiving nodes can validate their LSU packets. To do so, nodes either use a public key distribution packet specifically designed for this purpose or attach their keys to LSU packets. Key broadcasts are timed according to the network conditions and the device characteristics. For example, a node rebroadcasts its key when it detects substantial change in the zone topology. Nodes validate public key distribution packets only if they are receiving the public key of the originator for the first time. Upon validation, the public key E_v and the corresponding IP address are stored locally.

Each node commits its medium access control (MAC) and IP addresses (MAC_v, IP_v) to its neighbors by broadcasting signed 'Hello' messages. Receiving nodes validate the signature and update the neighbor table by retaining both MAC and IP addresses. The mappings in these two addresses stay in the table as long as transmissions from the corresponding neighboring nodes are overheard. A lost neighbor timeout period is associated with each table entry. When nothing is heard from a node for the timeout period, the entry is deleted from the table.

A node V broadcasts its link state data by using an LSU packet.

$$V \rightarrow \text{broadcast} : \{TYPE, R, Zone\_R, LSU\_Seq, LSU\_signature, Hops\_Traversed, LS\_Data\}$$

where *TYPE* is the packet type,

R is the number of hops from the node to the zone boundary,
$Zone\_R = H^R(X)$,
$Hops\_Traversed = H(X)$,
X is a random number,
H is the hash function that every node knows,
LSU_Seq is the sequence number of the LSU packet,
$LSU\_signature = \{TYPE, R, Zone\_Radius, LSU\_Sequence\}_{<D_V>}$.

Receiving nodes first validate the signature. If the LSU packet is valid, they can derive the link state information in the packet. Then they hash the Hops_Traversed value in the LSU packet.

$$Hops\_Traversed = H(Hops\_Traversed) \tag{12.9}$$

If the new *Hops_Traversed* value is equal to the *Zone_R* value after hashing, the packet has reached the boundary of the zone and should not be forwarded further.

12.3 Further Reading

There are several other secure routing protocols for ad hoc and sensor networks. Secure position-aided ad hoc routing (SPAAR) is one of them. SPAAR protects position information in a high-risk environment. In SPAAR, nodes verify their one-hop neighbors before including them in routing with the aid of position information. SIGF (Wood *et al.*, 2006) is another secure sensor network routing protocol that also relies on the location awareness of nodes.

In Parno *et al.* (2006) a sensor network routing scheme where security and efficiency are the central design parameters is introduced. The protocol incorporates all three security approaches, i.e. prevention, detection/recovery and resilience to attacks. On the other hand, the detecting and correcting malicious data (DCMD) scheme (Golle *et al.*, 2004) focuses on the detection of, and the recovery from, attacks.

In Buchegger and Le Boudec (2002), Capkun and Hubaux (2003) and Pan *et al.* (2007) other secure ad hoc network protocols are introduced. Fonseca and Festag (2006) provide an analysis and comparison of various secure ad hoc routing protocols.

12.4 Review Questions

12.1   How and why is the preventive approach more effective than the resilience approach in the design of secure ad hoc routing protocols?

12.2   What are the differences between temporal and geographical leashes? Which type is more effective?

12.3   Design a directional-antenna-based protocol that can also detect wormholes that replay data packets at the sector opposite to the one at which they are received. Discuss how practical and effective your design is.

12.4   How can a node ensure that all of its neighbors are physically separate nodes?

12.5   Does a positive or negative acknowledgement based scheme better detect a malicious node selectively forwarding? Why?

12.6   How is data aggregation in sensor networks related to multipath routing?

12.7   How can a random-walk-based routing scheme improve security? What are its weaknesses?

12.8   What does phantom flooding provide in addition to what baseline or probabilistic flooding techniques can offer?

12.9   How is spoofing prevented in the route discovery phase of INSENS?

12.10 Is an active-1-1 or active-0-10 attack more difficult to tackle? Why?

12.11 How does location information improve security in ad hoc routing? Does it also introduce additional challenges? If it does, is it still worth it?

12.12 How and why is TESLA used in secure routing protocols?

12.13 Do you think that ARIADNE or ARAN can also be used for sensor networks? Why?

12.14 How can a malicious node that introduces itself into a zone by broadcasting a public key distribution packet be detected in SLSP?

13

Specific Challenges and Solutions

In this chapter we explain other challenges specific to ad hoc and sensor networks. Instead of surveying all the solutions related to a specific challenge, one or two well-known schemes are elaborated for each challenge. Our goal is to introduce the specific challenges and their technical implications.

13.1 SPINS: Security Protocols for Sensor Networks

The first challenge is the stringent resource constraints, especially in wireless sensor networks. The hardware limitations for sensor networks are explained in Chapter 2. The memory available in a sensor node is typically in the order of Kbytes. Moreover, a sensor node has limited computational power and its lifetime is based on the lifetime of an onboard battery and therefore power consumption is always an important consideration. Finally, sensor nodes generally transmit short data packets, e.g. 30 bytes. This indicates that even a very low-cost security scheme may introduce high overhead, i.e. 3 bytes is 10% of a 30-byte packet.

All of these factors dictate the need for a very low cost security scheme. For example, asymmetric encryption schemes that require long signatures and overhead, e.g. 50–1000 bytes per packet, are impractical for sensor networks. Similarly, block ciphers, e.g. AES and DES, which require high computation codes or large look-up tables, may not be practical for sensor nodes. SPINS (Perrig *et al.*, 2001b) is perhaps the earliest security protocol that addresses these challenges. SPINS has two building blocks – namely the sensor network encryption protocol (SNEP) and μTESLA – to provide the following security services:

- **SNEP**

  - Data confidentiality
  - Authentication
  - Integrity
  - Freshness

- **μTESLA**

  - Authenticated broadcast

Security in Wireless Ad Hoc and Sensor Networks   Erdal Çayırcı and Chunming Rong
© 2009 John Wiley & Sons, Ltd

13.1.1 SNEP

SNEP uses a lightweight version of RC5 both for encrypting the messages and generating message authentication codes (MACs). Separate keys are used for encryption, i.e. K_{encr}, MAC generation, i.e. K_{mac}, and random number generation, i.e. K_{rand}. All these keys are bootstrapped from the initial master key shared by the base station and the node. A pseudo random function F is used to derive the keys. Since $F_K(X) = MAC(K, X)$ is the function to bootstrap the keys and the MAC has strong one-way properties, the keys produced are computationally independent, which means that if one key is compromised, both parties can derive a new key without transmitting any confidential information.

Another key feature of SNEP is a *counter* maintained by both the sender A and receiver B. After each successful transmission the counter is incremented by both A and B. The counter is to ensure freshness and semantic security. Its value is an input to the encryption. Therefore, even if the same plain text is transmitted twice, the encrypted text is different at each transmission, which provides semantic security for data confidentiality.

In SNEP, A sends the following message to B to transmit a data fragment D:

$$A \rightarrow B : \epsilon, M$$

where

ϵ is the encrypted data fragment, i.e. $\epsilon = \{D\}_{<K_{encr,\ c>}}$
M is the MAC, i.e. $M = MAC(K_{mac}, c|\epsilon)$
c is the counter value.

Note that the counter c and encrypted message ϵ are first concatenated, i.e. $c \mid \epsilon$, and given to the MAC generation function; the resulting MAC is transmitted together with the encrypted message ϵ. This provides data confidentiality and authentication. Since the counter c is used both for encryption and MAC generation, semantic security is also provided. The counter also ensures freshness. However, the counter used in plain SNEP can only ensure weak freshness, i.e. it ensures that the same message cannot be replayed by multiple nodes. For strong freshness, the sender is challenged by sending a nonce η, i.e. an unpredictable random number. To do this:

- node A generates a nonce η_A randomly and sends it along with a request message ρ_A.

$$A \rightarrow B : \eta_A, \rho_A$$

- node B returns the nonce η_A with a response message ρ_B after a MAC computation.

$$B \rightarrow A : \{\rho_B\}_{<K_{encr,\ c>}}, MAC(K_{mac}, \eta_A|c|\{\rho_B\}_{<K_{encr,\ c>}})$$

If the MAC verifies correctly, node A knows that node B generated the response after it sent the request. The first message can also use plain SNEP if confidentiality and data authentication are needed.

13.1.2 μTESLA

μTESLA is for authenticating broadcast messages. It is based on MAC, where each MAC key is a key from a key chain generated by a one-way function F, i.e. $K_i = F(K_{i+1})$. This means that if you know the previous key you can authenticate the next key but cannot generate it from the previous key. This is one of the key ideas in μTESLA.

The other important idea is to divide the time into key disclosure intervals t_i, and make all the nodes loosely time synchronized (Figure 13.1). During each time interval one key is used to generate a MAC for the broadcast packets, and the key used during the interval is disclosed at the end of it. Since the receivers know the previous key, they can first authenticate the key for that interval and then all the packets received in that interval if the key is authentic.

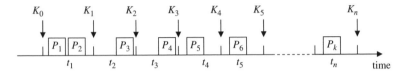

Figure 13.1   Key disclosure intervals in μTESLA

A broadcast period starts with a sender set-up session, where the sender generates a set of keys $K = \{K_0, K_1, K_2, \ldots, K_n\}$ starting from K_n. After this, loose time synchronization should be achieved and all the receivers need to be bootstrapped with K_0. To achieve this, the sender S, i.e. the node that broadcasts, is challenged by sending a nonce η. The scheme works as follows:

- The receiver node R generates a nonce η_A randomly and sends it to the sender S.

$$R \rightarrow S : \eta_A, \rho_A$$

- Sender S replies with a message containing its current time T_s, the key K_i used in the previous key disclosure interval, the starting time of the previous interval T_i, the length of a key disclosure interval T_{int} and the disclosure delay δ, i.e. the maximum delay to disclose a key at the end of a disclosure interval.

$$S \rightarrow R : T_s | K_i | T_i | T_{int} | \delta, \text{MAC}(K_{SR}, \eta_A | T_s | K_i | T_i | T_{int} | \delta)$$

The MAC in the reply uses the secret key shared by the sender and receiver. Since μTESLA is not for confidentiality but to authenticate a node that broadcasts, the sender does not need to encrypt the content of its reply.

13.2 Quarantine Region Scheme for Spam Attacks

The quarantine region scheme (QRS) introduced by Coskun *et al.* (2006) is another approach to reducing the cost of data security schemes. In SPINS, low-cost authentication and encryption schemes have been designed. Similarly, the quarantine region scheme uses a low-cost authentication scheme. Moreover, it tries to locate malicious nodes, called *antinodes*, and the nodes that are in the range of antinodes. The region that covers all the antinodes and the nodes within their range is called the *quarantine region*. Only the nodes in the quarantine region require authentication. This reduces the cost of authentication considerably, as well as preventing malicious nodes from running effective attacks to deplete the network resources (see the section DoS against Routing Schemes in Section 8.1.2.3).

In QRS, the quarantined set of nodes and quarantine regions are determined dynamically by using a distributed approach. Each sensor node decides whether it should be quarantined or not on its own by intermittently checking authentication failures in its transmission range. The duration of authentication checks is a random variable defined as a system parameter, as described in the rest of this subsection.

Not quarantined nodes have two different modes of operation in two alternating periods: (i) in *check the status periods*, nodes check spam activities; (ii) in *keep the status periods*, nodes do not perform any spam checks.

A sensor node does not relay unauthenticated messages during a *check the status* period t_c. If it receives an unauthenticated message during t_c, it first requests authentication from the last hop node of the message. If the last hop node fails on authentication, this indicates that the last hop node may be an antinode; therefore the node changes its status to quarantined. The notion of a quarantine region is exemplified in Figure 13.2. The quarantine region is basically an abstract region where the transmissions of the antinode may be received. This region has an amorphous shape due to the unpredictable propagation environment. The nodes in the quarantine region are those that capture antinodes' spam messages during their t_c.

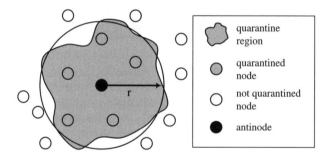

Figure 13.2 Boundaries of a quarantine region

An example sensor field is shown in Figure 13.3 where a quarantine region is indicated by the gray area. The nodes 3, 4, 7, 8 and an antinode are in the quarantine region and therefore they have to send and can relay only authenticated messages. Nodes outside the quarantine region do not need authentication to transmit a message, even if the message was an originally authenticated message coming from a quarantine region, unless their messages go through a quarantine region. For example, if node 11 receives an authenticated message from node 7 or 8, it transmits the message unauthenticated towards the sink because it is not in the quarantine region. On the other hand, if a node in a quarantine region (node 3 in the example) receives an unauthenticated message from a node outside of the quarantine region (node 1 or node 2), it first requests authentication from the corresponding node and relays the data packets only after successful authentication.

QRS is a dynamic scheme where sensor nodes periodically check the validity of their status. If a node does not detect an authentication failure during the *check the status* period t_c, its

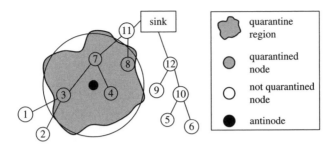

Figure 13.3   A sample sensor network and a quarantine region

status becomes *not quarantined*. The node stays in a *keep the status* period for t_k, and starts another *check the status* period after t_k. If the node detects any authentication failure in the *check the status* period t_c, it changes its status to *quarantined*. This is depicted at point *a* in Figure 13.4 where the node detects an authentication failure and starts a quarantined period. A node starts another *keep the status* period from the very beginning immediately it detects another authentication failure while in a *keep the status* period. The node exits quarantined mode and starts a *check the status* period only if it does not detect any authentication failures during a continuous *keep the status* period.

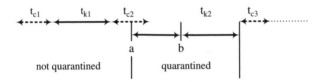

Figure 13.4   Timings for the quarantine periods

Both t_c and t_k (i.e. $t_c \geq t_{cmin}$, $t_k \geq t_{kmin}$, $t_c \in R^+$, $t_k \in R^+$) are random variables selected for each period, i.e. t_c and t_k may be different for every *check the status* and *keep the status* period. Sensor nodes determine these values according to the system-defined mean values, distribution functions and t_{cmin}, t_{kmin}. Since antinodes may be mobile or the propagation environment may change temporarily, this dynamic approach is needed.

This procedure and how it changes the location of a quarantine region dynamically is explained using the illustrative example in Figure 13.5. In this example, nodes *a*, *b*, *c*, *d* and *e* independently and asynchronously find out that there is a node within their transmission range that cannot be authenticated successfully by the end of their t_c, as shown in Figure 13.5(a). Therefore they become quarantined. Nodes *f* and *g* are not in the transmission range of the antinode; therefore they do not detect any authentication failure during their t_c, and become *not quarantined*. As shown in Figure 13.5(b), the antinode moves to a new location which changes its coverage area such that it includes the locations of nodes *f* and *g*. Since nodes *f* and *g* start a *check the status* period after every *keep the status* period t_k, they find out that they are in the coverage of an antinode in the first *check the status* period, and become quarantined. On the other hand, when nodes *a*, *b* and *c* do not detect any authentication failure for a whole *keep the status* period t_k, they change their status to *not quarantined*.

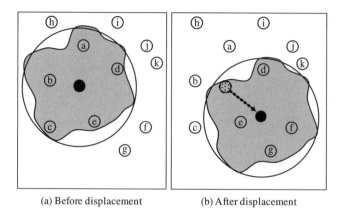

(a) Before displacement   (b) After displacement

Figure 13.5   The change in the boundaries of a quarantine region

Spam messages may get through a node during its *keep the status* period when it is not quarantined. *Local alarms* are used to minimize the effects of the spam messages that access the network at such times. A node that detects an authentication failure disseminates a *local alarm* to its *d*-hop neighbors by calling the broadcast_local_alarm function, where *d* is called the *local alarm depth*. As soon as a node receives a *local alarm*, it starts a *check the status* period. By using a local alarm mechanism, the nodes become more alert and the period during which an unsolicited message can get through from a node becomes limited.

Once a node sends a local alarm, it should not send another local alarm for a period of $k \times t_k$ even if it receives some other spam messages, where k is the local alarm factor, $k \in \mathbb{R}^+$. This is done in order not to flood the network with local alarms. An example is depicted in Figure 13.6, which shows the timings of local alarms by a node under attack. The node sends a local alarm at point *a*, where it detects an unsolicited message for the first time. Then it waits at least $k \times t_k$ before sending another local alarm. During this period it neither generates local alarms for the detected spam messages nor relays the local alarm messages of other nodes. For example, at points *b*, *c* and *d* it detects other unsolicited messages but does not send any local alarms for them. After the $k \times t_k$ period, it disseminates another local alarm after the detection of the unsolicited message at point *e*.

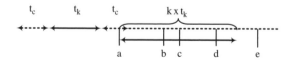

Figure 13.6   Timings of the local alarms

By using local alarms, buffer zones around quarantine regions are created, as shown in Figure 13.7. The nodes in a buffer zone are still *not quarantined* but are more alert than the nodes not quarantined. A node in the buffer zone runs the same algorithm as a *not quarantined* node. It just shortens each *keep the status* period by multiplying it by the *keep the status factor s*, where $0 < s < 1$. A node starts a *check the status* period as soon as it receives a local

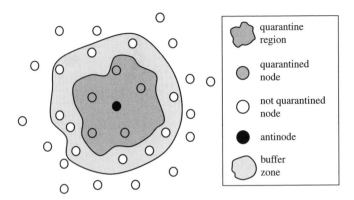

Figure 13.7 Buffer zones

alarm, which implies that the node is in the buffer zone of a quarantine region. If it does not receive another local alarm during $k \times t_k$, the node assumes that it is not in a buffer zone any more.

The QRS scheme can also tackle antinodes that have long transmission ranges. This kind of antinode can attempt to attack the entire sensor network from a distance. The QRS is independent of the location and the transmission range of the antinode. Since the antinode cannot be authenticated, its messages are not relayed by the sensor nodes, no matter where it is broadcasting its spam messages. In QRS, only the nodes that can receive the transmissions of the antinode require authentication, while the others do not.

13.3 Secure Charging and Rewarding Scheme

The charging and rewarding scheme targets nodes that deny the charges for services received from a multihop cellular network. This is called misbehaving with regard to the charging schemes and is examined in more detail in Section 8.1.2.4.

In Figure 13.8 an example scenario is depicted. Node A is the node that initiates communications with node B. The route from A to BS_A, i.e. the base station that the node A uses to access the infrastructure after multiple hops, is called the *upstream route*. The route from BS_B, i.e. the base station that the node B uses to access the infrastructure after multiple hops, to B is called the *downstream route*. The nodes that forward the data packets from A to BS_A are upstream forwarding nodes and the nodes between BS_B and B are downstream forwarding nodes. In our example we have one upstream and one downstream forwarding node, i.e. u and f respectively.

The charging and rewarding scheme (CRS) introduced by Salem *et al.* (2003) is based on the following approach:

- Authenticate the initiating node A and charge it for the communications before its packets are actually delivered. This is to prevent *refusal to pay* attacks.
- Identify the forwarding nodes and authenticate them. This is to ensure that only the selected nodes can forward and nodes that do not forward cannot claim that they do.
- Reward upstream nodes when the packets from A reach BS_A.

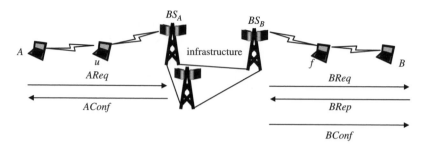

Figure 13.8   Session set-up for charging and rewarding in multihop cellular networks

- Reward downstream nodes when B acknowledges that the packets from A have been delivered to B.
- Charge B when the packets from A are forwarded to B by BS_B. Reimburse this charge when B acknowledges the delivery of packets, which is needed to enable the rewarding of downstream forwarding nodes.

13.3.1 Establishing a Session

A initiates the session set-up procedure by transmitting a set-up request $AReq$ in the following format:

$$A \rightarrow BS_A : AReq_0$$

$$AReq_0 = AReqID \mid oldASID \mid ARoute \mid TrafficInfo,$$

$$MAC(K_A, AReqID \mid oldASID \mid ARoute \mid TrafficInfo)$$

$AReqID$ is the identification of the request and is generated in sequence. In case the request is to re-establish a previously broken session, the identification of the broken session, i.e. $oldASID$, is also in the request. $OldASID$ is 0 if this is not a re-establishment effort. CRS assumes that there is a secure source routing protocol that can support the scheme. A obtains the route between A and BS_A from the secure source routing protocol, i.e. $ARoute$, and includes it in the $AReq$. Finally, information about the traffic to be sent, i.e. $TrafficInfo$, is also concatenated into the message. A also generates a MAC on the overall message by using its secret key K_A.

Every upstream forwarding node i checks $TrafficInfo$. If node i decides to forward the future packets in this connection, it computes a MAC on the request it received from the previous upstream forwarding node, i.e. $AReq_{i-1}$, by using its own secret key, replaces the MAC in the request and forwards $AReq_i$ to the next upstream forwarding node.

$$AReq_i = AReqID \mid oldASID \mid ARoute \mid TrafficInfo, MAC(K_i, AReq_{i-1})$$

Therefore, the request delivered to BS_A contains a single MAC computed by A and all the upstream forwarding nodes. BS_A repeats all the MAC computations and checks the result against the MAC in the received request. It also verifies that the $AReqID$ is fresh. This procedure authenticates all the nodes in the upstream route as well as A. Finally, if the $oldASID$

value is not 0, BS_A also checks if it is valid. If any of these verifications are not successful, BS_A drops the request. Otherwise, it sends the request to BS_B through the backbone.

When BS_B receives the request, it forwards it to the first downstream forwarding node to be sent to B:

$$BS_B \rightarrow B : BReq_0$$

$$BReq_0 = BReqID \mid oldBSID \mid BRoute \mid TrafficInfo$$

Note that $BReqID$ is a fresh identifier generated by BS_B. Like the upstream, every downstream node j also checks $TrafficInfo$ to decide if it wants to get involved in this session, generates a new MAC and replaces the MAC in the $BReq_{j-1}$ before forwarding it:

$$BReq_j = BReqID \mid oldBSID \mid BRoute \mid TrafficInfo, MAC(K_i, BReq_{j-1})$$

When B receives the request, and if it accepts this connection, it prepares the reply and sends it back. The reply contains only $BReqID$ and MAC, which was generated on the overall message forwarded to B by the last downstream forwarding node:

$$BReq_j = BReqID, MAC(K_B BReq_{B-1})$$

$BReq$ is conveyed back to BS_B through the same downstream route without any modification. BS_B generates the MAC for the route and verifies it with the returned MAC. If the verification does not fail, it informs BS_A. Then both BS_A and BS_B generate confirmation messages $AConf$ and $BConf$ respectively and send them to A and B:

$$AConf = AReqID \mid ASID \mid AMAC_A \mid AMAC_1 \mid \ldots \mid AMAC_a$$

$$AMAC_i = MAC(K_i, AReqID \mid ASID \mid oldASID \mid ARoute \mid TrafficInfo)$$

$$BConf = BReqID \mid BSID \mid BMAC_A \mid BMAC_1 \mid \ldots \mid BMAC_a$$

$$BMAC_j = MAC(K_j, BReqID \mid BSID \mid oldBSID \mid BRoute \mid TrafficInfo)$$

Each node i on the initiator route and j on the correspondent route verifies its own $AMAC_i$ and $BMAC_j$, and stores session identification $ASID$ or $BSID$ respectively. When confirmation messages are received, the session is set up.

13.3.2 Packet Delivery

In a packet delivery, source S may be the initiator A or the correspondent B node, and destination D may also be A or B interchangeably. In any case, A pays for the communications. When B sends a packet to A, it is charged temporarily and reimbursed as soon as A acknowledges. This is to prevent B from misbehaving and cheating A.

The ηth packet $SPkt_{0,\eta}$ sent by S contains the session identification $SSID$ ($ASID$ if S is A or $BSID$ if S is B), the sequence number η and the payload $Payload_\eta$ In addition to this, the message body includes a MAC computed on the packet using the secret key K_S:

$$SPkt_{0,\eta} = SSID \mid Body_{0,\eta}$$

$$Body_{0,\eta} = \eta \mid Payload_\eta \mid MAC(K_S, SSID \mid \eta \mid Payload_\eta)$$

The MAC in $SPkt_{0,\eta}$ can only be verified by BS_S. Therefore, all the upstream nodes simply ignore it. Instead, every upstream node i encrypts the body of the packet, including the MAC, by XORing it with $PAD_{i,\eta}$:

$$SPkt_{i,\eta} = SSID \mid Body_{i,\eta}$$

$$Body_{i,\eta} = PAD_{i,\eta} \oplus Body_{i-1,\eta}$$

PAD is generated by using a stream cipher. The session identifier $SSID$ and the secret key K_i are the seeds to initialize the stream cipher. $PAD_{i,\eta}$ is chosen as the ηth block of length $MaxLength$, which is the maximum allowed length of packets in bytes. If the actual length of the packet is smaller than $MaxLength$, unnecessary bytes are thrown away.

When the packet is received, the base station BS_S for the source node verifies the integrity of the packet and $SSID$. If any of the verifications fails, it drops the packet. Otherwise it forwards the packet to the base station BS_D for the destination node, which replaces $SSID$ with $DSID$, computes a new MAC, computes the $PAD_{j,\eta}$ for each downstream forwarding node j and XORs the packet with every $PAD_{j,\eta}$. Every downstream forwarding node j first decrypts the packet with its $PAD_{j,\eta}$ upon receipt and forwards to the next downstream forwarding node. Therefore, when a packet is delivered to D, all encryption PADs are removed.

13.3.3 Acknowledging Delivery

The destination D must acknowledge the receipt of packages. However, instead of acknowledging every packet, D sends a single acknowledgement batch in the following format when it considers the session to be closed.

$$DAck = DSID \mid Batch \mid LastPkt \mid LostPkts,$$

$$MAC(K_D, DSID \mid Batch \mid LastPkt \mid LostPkts)$$

To compute $Batch$, the $MAC(K_D, SSID \mid \eta \mid Payload_\eta)$ of every packet that will be acknowledged, i.e. all the packets except for the lost packets, are XORed. Note that $LastPkt$ is the sequence number of the last received packet and $LostPkts$ is the list of the sequence numbers of the packets that have not been delivered successfully.

13.3.4 Terminating a Session

Every node involved in the session starts a timer when the session is initiated. The timer is set every time a packet is forwarded for the session. When this timer expires, the node closes the session, which means it deletes the related state information from its memory. Another reason for closing a session is the detection of one of the following errors:

- a packet cannot be forwarded to the next hop;
- a packet with unknown session identification is received from a source node S;
- the node does not want to participate in packet forwarding any more.

13.4 Secure Node Localization

Node localization has a key role in many ad hoc network applications, and especially in wireless sensor networks. An attack on the node localization scheme may prevent an ad hoc network from fulfilling its expected functions. As we discussed in Chapter 8, there are many security attacks specifically designed to target localization schemes. There are several approaches to defending against these attacks:

- techniques against masquerading, replaying and node tampering also help to secure node localization schemes;
- secure routing techniques can defend against wormhole attacks designed to hamper node localization;
- multimodal localization schemes, where node localization is achieved based on multiple schemes such as both received signal strength indicator and time difference of arrival, can be used to estimate the distance to a beacon; if there is an inconsistency, this may indicate a malicious node;
- special techniques to assess the reliability of beacon nodes can be developed; they can be challenged, or the consistency of the data produced by them can be checked, and each beacon node can be assigned a value that indicates its level of reliability;
- consistency of the data coming from beacon nodes can be checked by using statistical methods and inconsistent data can be dropped;
- a node localization scheme can be designed robust enough to tolerate few false inputs.

In this section we will elaborate several schemes that follow some of these approaches.

13.4.1 Detection of Malicious Beacon Nodes and Replayed Beacon Signals

Nodes that generate false location information can be detected by the nodes that already know their locations – typically by the other beacon nodes. To do this, a beacon node n that receives a beacon signal from another beacon node n_a can estimate its location (x', y') according to the received beacon signal and compare the estimated location with its actual location (x, y). If the difference between these two locations is higher than a specified threshold τ, this may indicate that the node n_a that generates the beacon signal is malicious. In Liu *et al.* (2005a), this approach is followed for the case when the location estimation is based on a received signal strength indicator (RSSI) and beacon signals are unicast to the non-beacon nodes when non-beacon nodes request them. The scheme is comprised of the following steps:

- A beacon node n, i.e. the detecting beacon, requests a beacon signal, i.e. B_{req}, from another beacon node n_a, the target beacon node. The detecting beacon acts as though it is not a beacon node.

$$n \rightarrow n_a : B_{req}$$

- The target beacon sends the beacon signal, i.e. B_{beacon}. The beacon signal typically also includes the location (x_a, y_a) of the target beacon n_a.

$$n_a \to n : B_{beacon}$$

- The detecting beacon estimates the distance d_a to the location (x_a, y_a) of the target beacon based on the RSSI calculation.
- Since the detecting node already knows its location, it can calculate the distance d between itself and the target node location sent in B_{beacon}. If the difference between the estimated distance d_a, and the calculated distance d is higher than the threshold τ, this may indicate that the target node is malicious.

$$\text{if } \left| \sqrt{(x - x_a)^2 + (y - y_a)^2} - d_a \right| > \tau, \text{ then target is malicious}$$

Although this technique can detect anomalies in the beacon signals, beacon signals may be replayed by the other nodes, and therefore they may not always indicate that the beacon is malicious. This is especially true in the case of wormhole attacks. To tackle this issue, replayed beacon signals should be filtered out. Techniques to thwart wormhole attacks, explained in the previous chapter, can eliminate the beacon signals forwarded through wormholes. However, they cannot detect locally replayed beacon signals, i.e. a malicious node neighbor to a beacon blocking its signals and replaying them as if it is the beacon node.

In Liu *et al.* (2005a) a round trip time (RTT) based scheme is introduced. In this scheme, the maximum *expected RTT* between two nodes is tested and known before deployment. The nodes challenge the beacons for a beacon signal and also calculate the RTT for the beacon signal using Equation (13.1). If the *calculated RTT* is higher than the *expected RTT*, this indicates a locally replayed beacon signal, and therefore it is dropped.

$$RTT = (t_4 - t_1) - (t_3 - t_2) \tag{13.1}$$

In Equation (13.1) t_1 is the time at which the sender finishes sending its first byte and t_4 is the time at which it finishes receiving the first byte of the reply. Basically, $(t_4 - t_1)$ is the difference between the times that the sender sends the request and receives the reply. Both t_4 and t_1 are available at the sender. However $(t_4 - t_1)$ also includes the processing time at the replier as well as the delays due to MAC. Therefore, this value may vary based on many parameters. To clear these unpredictable factors from RTT, $(t_3 - t_2)$ is subtracted from it. $(t_3 - t_2)$ is the time between the receiver receiving the first byte of the request t_2 and it finishing sending the first byte of its reply t_3.

Detecting malicious beacons does not suffice to thwart their attacks. A malicious node can claim that beacons are adversaries. Therefore, a scheme is needed to ensure that the detecting beacons are friendly and their reports are correct. In Liu *et al.* (2005a) the authentication scheme between the beacon nodes and the base station is relied on for this purpose. Beacons report their findings to the base station. The base station maintains a *report* and an *alert counter* for each beacon. Beacons are not allowed to report at a rate higher than a specified value. The report counter is used for this. The alert counter is the reliability value for a beacon. Every time a negative report is received for a beacon, its alert counter is incremented. When the alert counter is over the threshold, the beacon is revoked.

13.4.2 Attack-Resistant Location Estimation

Inconsistency among the location data can be detected by inspecting the mean square error of estimation (MMSE) given by

$$\varepsilon = \frac{\sum\limits_{i=1}^{m} \left(d_i - \sqrt{(x - x_i)^2 + (y - y_i)^2} \right)^2}{m} \tag{13.2}$$

where

ε is the mean square error,
(x_i, y_i) is the location of beacon node i,
(x, y) is the estimated location,
d_i is the distance to beacon node i,
m is the number of beacon nodes used in the location estimation.

We can derive (x, y) by minimizing ε when the (x_i, y_i, d_i) triplet is known for at least three beacon nodes n_i, i.e trilateration. Since there will always be an error in estimating d_i, ε never becomes 0 in practice, but it should be below an acceptable level if all the nodes are benign. In Liu *et al.* (2005b) a scheme to filter out false data from the location estimation is proposed based on this observation. The mean square error ε is called the *inconsistency indicator*. When the inconsistency indicator ε is above a threshold τ, this indicates that there is a malicious beacon node introducing false data. The objective of the scheme is to find out the greatest set of beacon nodes that can be used to estimate the location such that the inconsistency indicator ε is below the threshold τ. A greedy approach is presented for this. The algorithm first starts with all beacon signals received from n beacon nodes. If the inconsistency indicator ε is above the threshold τ, the combinations with $n - 1$ beacon nodes are tried. The first combination that returns an inconsistency indicator ε below the threshold τ is picked as the solution. If there is no solution for $n - 1$ beacon nodes, this may indicate more than one malicious beacon node, and the $n - 2$ beacon combinations are checked. This continues until a solution where the inconsistency indicator ε is below the threshold τ is found or there is no solution for three or more beacon node combinations. The latter case indicates that the location cannot be estimated by using the available beacon signals.

Determining threshold τ is an important issue for this algorithm. If it is too high, malicious beacon signals may not be detected. On the other hand, when it is too low some benign beacon signals may be considered false and filtered out. In Liu *et al.* (2005b) the distribution for the inconsistency indicator is derived through simulation and analytical study, and this is used to determine the threshold. Field tests can also be carried out to derive the distribution for the threshold.

A similar robust statistical method is also introduced in Li *et al.* (2005). We leave that as a further reading exercise for the reader.

All these statistical approaches assume that an error below a certain threshold is acceptable, and this error is dependent on the number of beacon nodes and the average distance between the beacon nodes and the other nodes. Another attack-resistant localization technique based on the same assumption is called *voting-based location estimation* (Liu *et al.*, 2005b) where

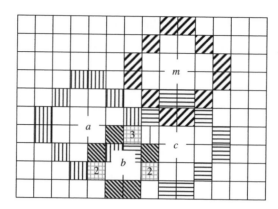

Figure 13.9 Voting scheme for localization

the field is partitioned into smaller square cells and each cell is given a value based on beacon signals, as shown in Figure 13.9. In a voting-based scheme the value of a cell that is a distance between $d_i - e$ and $d_i + e$ from a beacon node i is incremented where d_i is the estimated distance from the beacon node i and e is the acceptable error. The distance d_i is derived from the beacon signal from node i. After this is done for each received beacon signal, the center of the cell that has the highest value is the location. For example, in Figure 13.9, we have four beacon signals from nodes a, b, c and m. Nodes a, b and c are benign. When the voting scheme is applied based on the beacon signals by them, one of the cells has three votes, two cells have two votes and all the other cells have either one or zero votes. Applying the beacon signal by malicious node m does not change this result. The center of the cell that has three votes is the location.

13.5 Secure Time Synchronization

Security threats to time synchronization, as well as the defense mechanisms against them, are similar to the threats and counter measures in node localization. In this section we explain several techniques to detect replay attacks against synchronization schemes. A malicious node can block a synchronization message and replay it later. This is called a replay attack and may have a detrimental effect on the time synchronization. The technique presented in this section was introduced in Ganeriwal *et al.* (2005) and is based on observations of RTT like the technique against replay attacks in node localization (Liu *et al.*, 2005a). Note that this technique does not guarantee time synchronization under a replay attack, but detects replay attacks. It consists of three steps (Ganeriwal *et al.*, 2005):

- **Step 1:** $A(t_1) \rightarrow (t_2)B : A, B, N_A$, synch
  Node A sends node B a synchronization message at t_1, and the message is received by node B at t_2. The message includes the identifications of nodes A and B, a nonce N_A, i.e. a random number generated by node A and a timestamp *synch*. A nonce is used to ensure the long-term freshness of the packet, as explained in previous sections. It prevents a malicious node from using a MAC eavesdropped from a synchronization message.

- **Step 2:** $B(t_3) \rightarrow (t_4)A : B, A, N_A, t_2, t_3, ack, MAC(K_{AB}, B \mid A \mid N_A \mid t_2 \mid t_3 \mid ack)$
  Node B replies to node A at t_3, and the reply message is received by node A at t_4. Both t_1 and t_4 are already known to node A. Therefore, node B informs node A about t_2 and t_3. N_A is also in the reply message, which also serves to identify that the message has been replied to. A MAC produced from the reply message and the secret key K_{AB} between nodes A and B is also appended to the message.
- **Step 3:** if $(t_4 - t_1) - (t_3 - t_1) < \theta$, proceed. Otherwise abort.
  Node A calculates RTT. If RTT is smaller than the maximum RTT threshold, the synchronization is accomplished. Otherwise it is aborted.

Similar security schemes developed for the multihop and group sender/receiver synchronization cases appear in Ganeriwal *et al.* (2005).

13.6 Secure Event and Event Boundary Detection

For many applications the ultimate goal of a sensor network is to detect a set of predefined events or targets and to classify them. Sensor networking technologies enable the deployment of many sensor nodes such that multiple nodes can detect the same event. Therefore, they rely on the collaborative effort of many nodes to detect an event.

An event actually does not occur at a single point but affects a space. For example, a chemical attack may contaminate a large area. The boundary of such an area is called the *event boundary*, and the detection of the boundary is also an important task for sensor networks. This task may become more challenging in the presence of malicious or faulty nodes.

An adversary may try to inject false data to prevent a sensor network detecting events or event boundaries correctly. Conventional techniques, such as authentication and encryption, can be used to defend against such attacks. An alternative or complementary technique is based on the fact that there are other nodes in the vicinity of every node that can detect the same event, and nodes in a sensor network can collaboratively work to filter out false event data.

In Ding *et al.* (2005) an event boundary detection scheme that can detect event boundaries correctly when as many as 20% of the nodes are faulty, i.e. malicious or malfunctioning, is introduced. They assume N nodes are uniformly deployed in a two-dimensional Euclidean plane R^2. An event, denoted by E, is a subset of R^2 such that the readings of nodes in E are significantly different from those nodes not in E. A point $x \in R^2$ is said to be in the boundary of E when any disk centered at x contains both points in E and points not in E. The scheme has three stages, as outlined below.

13.6.1 Stage 1: Detection of Faulty Nodes

The fundamental idea in Ding *et al.* (2005) is that a measurement by a sensor should be correlated with the events in its vicinity, and those events are also detected by the other nodes within the event boundaries. Therefore, they compare the readings of a sensor S_i with the readings of the sensors in the set $N(S_i)$ that represents a close neighborhood of the sensor S_i. Let x_i denote the reading of the sensor S_i, and $x_{i1}, x_{i2}, \ldots, x_{ik}$ denote the readings of the other nodes in $N(S_i)$. Then, the difference d_i between x_i and the center of $\{x_{i1}, x_{i2}, \ldots, x_{ik}\}$ is

$$d_i = x_i - med_i \qquad (13.3)$$

In Equation (13.3), med_i is the median of $\{x_{i1}, x_{i2}, \ldots, x_{ik}\}$ but not the mean, i.e. not $(x_{i1} + x_{i2} + \ldots + x_{ik})/k$. The median is preferred because it is a robust estimator of the center of a sample and filters out extreme values.

The next step examines whether this difference d_i for sensor S_i is normal or too much compared to the difference d values for the other $n - 1$ sensors around S_i. To do this, another closed bounded set of sensors $N^*(S_i) \subset R^2$ around sensor S_i, which is typically larger than $N(S_i)$ but can also be equal to $N(S_i)$ is selected. $N(S_i)$ is for calculating d_i while $N^*(S_i)$ is for calculating the mean μ and standard deviation σ of set $D = \{d_1, \ldots, d_i, \ldots, d_n\}$ around sensor S_i. Then d values are calculated for every sensor in $N^*(S_i)$. For example, in Figure 13.10 $N(S_1)$, $N(S_i)$ and $N(S_n)$ are used to calculate d_1, d_i, and d_n respectively for sensors S_1, S_i and S_n, which are the members of $N^*(S_i)$. Note that:

$$N(S_i) \subset N^*(S_i)$$

$$N^*(S_i) \subset (N(S_1) \cup N(S_i) \cup N(S_n))$$

$$N^*(S_i) = \{S_1, \ldots, S_i, \ldots, S_n\}$$

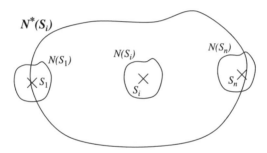

Figure 13.10   The neighborhood sets for calculating the difference

The mean μ and standard deviation σ of set $D = \{d_1, \ldots, d_i, \ldots, d_n\}$ are given by

$$\mu = \frac{1}{n} \sum_{i=1}^{n} d_i \tag{13.4}$$

$$\sigma = \sqrt{\frac{1}{n-1} \sum_{i=1}^{n} (d_i - \mu)^2} \tag{13.5}$$

Now it is possible to standardize the difference values of sensors according to the difference values of the other nodes in their vicinity by

$$y_i = \frac{d_i - \mu}{\sigma} \tag{13.6}$$

The standardized difference value y_i is a good indicator of whether sensor S_i generates faulty readings or not. If $|y_i| \geq \theta$, where $\theta > 1$ is a predetermined threshold, then we can conclude that S_i is faulty. Again, the threshold θ is an important value. If it is too low, correct readings can be accepted as faulty. On the other hand, if it is too high, faulty messages may not be detected.

13.6.2 Stage 2: Detection of Event Boundary Nodes

Let Ω_1 denote the set of faulty nodes found in the first stage. In the second stage the nodes at the event boundary are picked from among the nodes not in Ω_1, i.e. $S - \Omega_1$. To do this, the difference value d_i of a node S_i is recalculated after decomposing $N(S_i)$ into smaller neighborhood areas $NN(S_i)$. For an event boundary node, the difference value for the sub areas that are largely outside the event region E is higher than for the sub areas in the event region E. This observation is used to locate the nodes at the boundary.

For example, in Figure 13.11 the disk that represents $N(S_i)$ is divided into three sectors of $120°$. Let's assume that S_i is a node at the event boundary, which means S_i detects the event, i.e. it is in the event region E, but many nodes in $N(S_i)$ cannot detect the event because they are not in E. Therefore, the differences d_{iA} and d_{iB} for Sectors A and B would be higher compared to d_{iC} for Sector C because all the nodes in Sector C are detecting the event just like S_i, but many nodes in Sectors A and B are not.

The algorithm introduced to detect the boundary nodes in Ding *et al.* (2005) can be summarized as follows:

1. Construct the set of faulty nodes Ω_1.by running the algorithm in Stage 1.
2. For each sensor S_i not in Ω_1, i.e. $S_i \in S - \Omega_1$, perform the following steps:

   (i) Partition the $N(S_i)$ into sectors.
   (ii) Calculate the difference d_{ij} for each sector j.
   (iii) Assign the largest d_{ij} as the new d_i for S_i.
   (iv) Without changing the difference values of the other nodes, recalculate the mean μ, standard deviation σ and y_i for $N^*(S_i) - \Omega_1$ and the new d_i.
   (v) If $|y_i| \geq \theta_2$ after recalculation, S_i goes into the set of boundary nodes denoted by Ω_2.

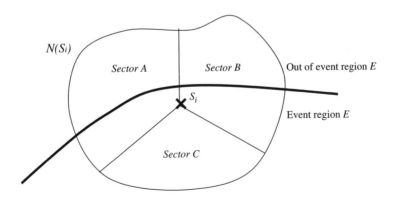

Figure 13.11   Partitioning of a neighborhood disk

13.6.3 Stage 3: Improvement of Event Boundary Node Detection

The standardized difference value y_i of a node S_i at an event boundary is normally higher than the nodes inside or outside an event. Therefore, some nodes in Ω_1, i.e. the nodes found faulty at the end of Stage 1, may be at the event boundary. Similarly, some nodes in Ω_2, i.e. the nodes found to be boundary nodes at the end of Stage 2, may not actually be at the event boundary. Stage 3 reduces this kind of false derivation. The algorithm is based on the assumption that the nodes are uniformly distributed and therefore within a certain distance c from a boundary node there should be at least one other boundary node. The algorithm can be summarized as follows:

- For each node S_i in $\Omega_1 \cup \Omega_2$
- If there is at least one other node S_j, i.e. not S_i, in $\Omega_1 \cup \Omega_2$ which is closer than c to S_i, then S_i is in Ω_3.

Here Ω_3. is the set of nodes at the event boundary. Note that in this algorithm c is an important parameter, similar to θ and θ_2, and should be selected very carefully.

13.7 Review Questions

13.1 What kind of security services does SNEP provide? How does SNEP differ from the conventional schemes that provide the same security services?

13.2 What are the differences between TESLA and μTESLA?

13.3 Assume that nodes A, B and C are in a quarantine region and nodes D, E and F are not. Which of the following transmissions need to be authenticated?

(a) $B \rightarrow A$
(b) $D \rightarrow B$
(c) $D \rightarrow A$ and then $A \rightarrow E$
(d) $C \rightarrow F$
(e) $E \rightarrow F$

13.4 When can we charge a node which is the source of a packet in a hybrid cellular and ad hoc network? Why?

13.5 How can ad hoc nodes be prevented from cheating the system by claiming they forwarded packets on behalf of others even though they did not?

13.6 Explain two approaches to detecting malicious beacon nodes?

13.7 How can round trip time be used to detect replay attacks in node localization and time synchronization?

13.8 The difference values d_i for nodes S_i are $\{5, 7, 9, 2, 3, 6, 4, 11, 5, 6\}$ respectively. If the threshold is 1.5, which nodes are faulty?

14

Information Operations and Electronic Warfare

We would like to start this chapter with a disclaimer: 'Definitions, views and opinions expressed in this chapter do not necessarily reflect the official point of view of any national or international organization.' We compiled the content of this very short chapter from open and unclassified material available on the Internet. Our goal is to provide a brief survey of the related terminology. We do not intend to provide an extensive survey where all the challenges and solutions are explained.

Ad hoc and sensor network applications are already employed for tactical systems such as friendly force tracking, unattended sensor networks, local area communications, maritime and land surveillance/reconnaissance. Tactical communications is naturally an ad hoc networking application. Concepts like network-centric warfare make ad hoc networking capabilities more and more important for the military. Therefore, the tendency is to increase the usage of these technologies in defense systems, which are targets for information operations. In addition to this, due to the effect-based operations approach and other state-of-the-art concepts, civilian systems that connect important nodes in national infrastructures and that are used in critical infrastructures are primary targets for adversaries in a war. Moreover, due to asymmetric warfare and low-intensity conflicts, the threats against information systems exist during crises and peace periods (Hammes, 2007). Hence, civilian systems are also subject to threats and the threats are continuous.

Today, the Internet and other means of internetworking tie important national assets to each other as well as to global resources. The public also has access to 24-hour informative and/or deceptive news channels. All these increase the vulnerabilities of global and national societies. Now, economic and diplomatic pillars of national power can be dragged into a chaotic, uncertain and weak status by the use of information operations much easier than in the era before the Internet. The notions of ubiquitous, sentient and pervasive computing, and their widespread use, further increase the risks. When they are not protected, banking systems, stock exchange systems, smart offices and homes, patient and elderly care systems, surveillance systems,

Security in Wireless Ad Hoc and Sensor Networks   Erdal Çayırcı and Chunming Rong
© 2009 John Wiley & Sons, Ltd

habitat monitoring systems, automated critical infrastructures, automated transportation networks and infrastructures and many other systems that use these technologies can be targets for asymmetric threats (i.e. terrorist acts), which not only cause financial losses but also casualties. By the use of automated systems, sentient and ubiquitous computing tools, it may be possible to trap crowds, to cause trains or aircraft to crash, to start fires or to flood rooms with chemicals. Therefore, the authors believe that information operations can also be lethal and destructive even without using conventional destructive means, if carefully coordinated and integrated measures are not taken and exercised. Note that this view about lethality is different from the common current perception about information and electronic warfare. All these may also have an impact on public morale, which in turn can weaken military power and its effective usage.

The bottom line is that information operations are continuous even in peacetime and require both civilian and military attention. Information operations can be conducted offensively or defensively, and aim:

- to prevent adversaries from using their information systems and command and control (C2) infrastructure effectively;
- to gain intelligence;
- to use the information and automation systems owned by the adversary to organize nonlethal or destructive attacks;
- to protect one's own interests, information, information systems, information-based operations and C2 infrastructure against the offensive information operations of adversaries.

In military terms this requires the usage of the following capabilities:

- **Electronic warfare (EW):** controlling and securing the friendly use of the electromagnetic spectrum and denying its use by the adversary.
- **Operational security:** securing friendly operations such that the adversary cannot learn friendly capabilities or future plans.
- **Psychological operations:** activities planned and conducted to influence the attitudes and behavior of friendly, neutral or enemy audiences such that they support friendly operations.
- **Military deception:** used to mislead the adversary into a course of action in favor of the friendly interest. This can be done by manipulation, distortion and/or falsification of evidence and information.
- **Physical destruction:** destroying the opponent's information and C2 systems by means of naval, air and land operations.
- **Special operations:** in the information operations context, operations planned and conducted by special operations units to destroy or disrupt the enemy's information and C2 systems as well as to gain access to the enemy's information systems. Special information operations may require long-distance infiltration into enemy territory by the special operations units to conduct electronic warfare, operational security, psychological operations, military deception and physical destruction tasks.

These capabilities should be planned and used in a well-coordinated and integrated manner, as well as being mutually supported by all means of communication, command, control, computer, intelligence, surveillance and reconnaissance (*C4ISR*) systems. All these capabilities are in the scope of information operations. However, we explain only electronic warfare in more detail in this chapter because this subject is more related to security in wireless ad hoc, sensor and mesh networks. Note again that our intention is not to provide extensive coverage of this topic. We give only the definitions related to electronic warfare. Readers more interested in this subject can refer to Joint Chiefs of Staff (2007).

Electronic warfare is a nonlethal but significant capability. It may impact the results of operations in all kinds of conflicts. The full electromagnetic spectrum is within the scope of electronic warfare, which includes activities that can be categorized into three broad classes:

1. **Electronic support (ES):** ES measures (ESMs) consist of mainly passive techniques to search for, intercept, identify and locate the source of an electromagnetic emission. They generally provide time-sensitive intelligence.
2. **Electronic attack (EA):** EA includes electronic counter measures (ECMs) to prevent an adversary from using the electromagnetic spectrum and C2 systems effectively.
3. **Electronic protection (EP):** EP involves measures (EPMs) to defend against ESMs and ECMs.

14.1 Electronic Support

ES includes *signal intelligence* (SIGINT) activities, which can be categorized into two classes: *communications intelligence* (COMINT) and *electronic intelligence* (ELINT). COMINT is for detecting and intercepting the adversary's communications. This may be in the form of traffic analysis to monitor which nodes are communicating and/or in the form of eavesdropping to capture the contents of messages. ELINT is conducted to detect and locate the source of an electromagnetic emission not used for communications, e.g. a radar emission.

COMINT is typically conducted in two phases. *Signal collection* is the first phase where the communications are recorded for the second phase – *traffic selection*. Selecting a data packet or a voice record that has intelligence value may be a very difficult task when it needs to be done among a very large amount of signals collected. Signal collection is carried out by tapping and eavesdropping and can be considered easier than traffic selection. Therefore, marking data packets that contain classified information may not be a good idea. Do you remember? The optional security header in IPv4 is for marking data packets that contain classified data. This may be helpful – but more so to adversaries!

RF fingerprinting is an important ES technique. The signals from two stations that are communicating at the same central frequency by using the same bandwidth and protocol suite normally have some differences due to unintentional frequency modulation, harmonics, etc. These differences can be detected by careful signal analysis and a radio can be distinguished from the others. This makes anonymity harder to achieve. When node localization schemes are used together with RF fingerprinting, the movement of nodes can be monitored, which may reveal critical tactical intelligence.

We examine node localization from the node owner's point of view in Chapter 7. However, triangulation or multilateration techniques used for locating one's own nodes may also help an adversary to locate them. Therefore, node localization schemes developed for tactical systems should also consider an enemy's SIGINT activities.

Finally, *threat warning* is also an important ESM. Electromagnetic emissions can indicate and locate enemy targeting, intelligence, tracking and guidance systems. Detection of such threats and consequent warning of friendly forces and troops can help to deploy electronic counter measures on time. For example, a pilot can be warned about a missile tracking his/her aircraft.

14.2 Electronic Attack

EA includes both offensive and defensive ECMs. Offensive ECMs, such as those listed below, are carried out to support operations by degrading the enemy's efficient use of the electromagnetic spectrum:

- **Jamming:** jamming and other denial-of-service attacks have already been covered in Chapter 8. In EW terminology they are also called *soft kill techniques*. Jamming can be used for offensive purposes, such as disturbing enemy communications to support a friendly attack.
- **Physical destruction:** jamming temporarily disrupts an adversary's transmission or emission at a certain bandwidth. On the other hand, links or nodes can be physically destroyed, which is also called *hard kill*.
- **Anti-radiation missiles:** guidance systems that lead ammunition towards the source of an emission are available. They are often used for suppressing enemy radar-guided weapon systems, such as anti-aircraft artillery guns.
- **Electronic deception:** deception is synonymous with masquerading, replay and spoofing attacks. False emissions are made to convey misleading information to the enemy. Intentional dummy transmissions for deceiving or confusing the enemy are called *electromagnetic intrusion*. Another approach to electronic deception is called *electronic probing*, where signals are sent to the adversary's devices to learn their functions or capabilities.
- **Directed energy (DE), high energy RF (HERF):** these are systems that emit large currents to damage unprotected electronic circuitry. Electromagnetic pulse devices that can create intentional damage by the use of electromagnetic emissions are being developed.

Defensive ECMs are conducted to protect friendly forces, and include:

- **Jamming:** jamming can also be used for defensive purposes. For example, enemy radar can be jammed to protect friendly forces from radar-guided missiles.
- **Expendables (chaff, flare and decoy):** expendables are also defensive jamming techniques. Chaff is used for jamming radar signals and is made up of thin strips of conducting foil that are half the wavelength of the target signal in size. They are dispersed between the radar and its target to cause false returns of radar signals. Similarly, decoys can be inserted between the radar and its targets to shield them. Flares can jam guidance systems that track a heat source.
- **Counter radio-controlled improvised explosive devices (RCIEDs):** improvised explosive devices are often remote controlled. Jamming these remote controls without accidentally triggering them is an important electronic counter measure.

14.3 Electronic Protection

Electronic protection (EP) consists of all measures against enemy ESMs, such as careful usage of the electromagnetic spectrum, authentication and cryptography, as well as all electronic counter counter measures (ECCMs) to defend against the enemy's ECMs, such as hardening equipment to resist directed and high-energy attacks and the destruction of enemy jammers. All the techniques for traffic analysis, anonymity and authentication that we explain in this book are protective measures against enemy ESMs. EP also includes:

- **Emission control (EMCON):** defined as controlled use of the electromagnetic spectrum. The probability that an emission is detected by enemy sensors is minimized by EMCON. Emissions should also comply with the deception plans.
- **Spectrum management:** also important for the effective utilization of the electromagnetic spectrum. Note again that spectrum is a precious resource and friendly emissions can also interfere with one's own electronic devices. Moreover, in a battlefield, spectrum is also subject to an adversary's ECMs. Therefore, its usage needs to be planned and managed carefully. The procedures for spectrum management should also be cognitive and adaptive to enemy ECMs.
- **Electronic masking:** for covering friendly emissions by jamming enemy ESMs. Controlled radiation on friendly frequencies can mask one's own emissions from enemy ESMs without interfering with them. For example, directional antennae can be used for this purpose.
- **Electronics security:** includes the techniques to deny ELINT.
- **Electromagnetic hardening:** a counter measure against ECMs. By filtering, grounding, bonding and shielding, personnel and equipment can be protected from enemy EA.

Chapter 8 in this book can be considered the section about EA, and all the following chapters are about EP. Of course, the scope of this book is wider than EW. It also includes information operations.

14.4 Review Questions

14.1 Relate the following concepts with their counterpart EW terminology explained in this chapter:

(a) Tamper resilience
(b) Sybil attack
(c) Integrity, confidentiality and privacy
(d) Masquerading, spoofing, phishing
(e) TESLA
(f) Secure node localization and time synchronization
(g) Secure event boundary detection

15

Standards

15.1 X.800 and RFC 2828

ITU-T Recommendation X.800 (Security Architecture for OSI) and IETF RFC 2828 (Internet Security Glossary) are used as references to systematically evaluate and define security requirements. Though coming from different standardization bodies, the two standards have many points in common. X.800 is used to define general security-related architectural elements needed when protection of communication between open systems is required. X.800 establishes guidelines and constraints to improve existing recommendations and/or to develop new recommendations in the context of OSI. Similarly, RFC 2828 provides abbreviations, explanations and recommendations for information system security terminology.

Both X.800 and RFC 2828 are designed to assist security managers in defining security requirements and possible approaches to meeting those requirements. They also help hardware and software manufacturers to develop security features for their products and services that follow certain standards. X.800 and RFC 2828 both mention several aspects of security systems, namely security threat and attack, security services and mechanisms and security management. This section gives a brief introduction to these standards. We urge readers to read the original standard documents for more information.

15.1.1 Security Threats and Attacks

According to X.800, 'A threat to a system security includes any of the following: destruction of information and/or other resources; corruption or modification of information; theft, removal or loss of information and/or other resources; disclosure of information and interruption of services'. Another, clearer definition comes from RFC 2828, which defines a threat as 'A potential violation of security exists when there is a circumstance, capability, action, or event that could breach security and cause harm'. In other words, a threat is a possible danger that might exploit vulnerability.

Security in Wireless Ad Hoc and Sensor Networks   Erdal Çayırcı and Chunming Rong
© 2009 John Wiley & Sons, Ltd

Threats can be classified as accidental or intentional and may be active or passive:

- Accidental vs. intentional threats – as their names imply, accidental threats exist with no premeditated intent; for example, system malfunctions or software bugs. On the other hand, intentional threats are planned actions for specific purposes.
- Passive vs. active threats – passive threats do not modify the information in or operations of the victim systems; for example, wire tapping. Active threats, on the other hand, involve modification of information in or operation of the victim systems; for example, changing the firewall rules of a system to allow unauthorized access.

While a threat is a potential security problem that may lead to a security breach, it is not yet an action. An attack, on the other hand, is an action to exploit a security breach. Attacks can also be classified as insider or outsider attacks, and active or passive attacks:

- Insider vs. outsider attacks – insider attacks occur when legitimate users of a system behave in unintended ways. Outsider attacks are initiated from outside the security perimeter by illegitimate system users.
- Active vs. passive attacks – active attacks attempt to change system resources or affect their operation. Examples of active attacks are masquerade, replay, modification of message and denial of service. Passive attacks attempt to make use of information from the system without changing system resources. Examples of passive attacks are message content disclosure and traffic analysis.

15.1.2 Security Services

According to RFC 2828, a security service is a processing or communication service provided by a system to protect system resources. Security services implement security policies and are implemented by security mechanisms. Security services are divided into five categories:

- **Authentication service:** this security service verifies the identities claimed by or for an entity (cf. p.16, RFC 2828). Authentication services are divided into two groups: data origin authentication and peer entity authentication.

  - *Data origin authentication*: this security service verifies the identity of a system entity that is claimed to be the original source of received data (cf. p.53, RFC 2828). It does not provide protection against duplication or modification of data units even though it is sometimes thought to enable a recipient to verify that the data have not been tampered with in transit.
  - *Peer entity authentication:* this service provides corroboration between peer entities at the connection establishment or during the transfer of information between them. This service guarantees that an entity is not attempting to masquerade or to replay a previous connection without authority (cf. p.8, Recommendation X.800).

- **Access control:** this service provides protection against unauthorized use of resources such as computing resources, storage resources, communication links, etc. To use a resource, the user should first be authenticated, after which they can be granted the right to use specific system resources.

- **Data confidentiality:** this service protects data from unauthorized disclosure as the data are transmitted from a source to a destination. Encryption and decryption are often used to provide data confidentiality. Data confidentiality is divided into four groups:

  - *Connection confidentiality:* this service provides confidentiality of user data on a connection.
  - *Connectionless confidentiality:* this service provides confidentiality of user data for connectionless services, i.e. it protects individual data blocks.
  - *Selective field confidentiality:* this service provides confidentiality of selected fields of user data in a connection or in an individual data block.
  - *Traffic flow confidentiality:* this service protects information which might be derived from the observation of traffic flows. It serves to protect against traffic analysis.

- **Data integrity:** this service ensures that the data are received exactly as they were sent and there has been no modification or replay of the data. Data integrity is classified into five groups (cf. pp. 9–10, Recommendation X.800):

  - *Connection integrity with recovery:* this service provides integrity for all user data on a connection, detects any modifications, insertions, deletions or replays of any data within an entire data sequence and attempts to recover the data if an attack is detected.
  - *Connection integrity without recovery:* this service provides integrity for all user data on a connection and detects any modifications, insertions, deletions or replays of any data within an entire data sequence but does not attempt to recover the data when an attack is detected.
  - *Selective field connection integrity:* this service provides integrity for selected fields within the user data transferred over a connection and takes the form of determination of whether the selected fields have been modified, inserted, deleted or replayed.
  - *Connectionless integrity:* this service provides integrity for individual data blocks and may take the form of determination of whether a received data block has been modified. Additionally, a limited form of detection of replay may be provided.
  - *Selective field connectionless integrity:* this service provides integrity for selected fields within individual data blocks and takes the form of determination of whether the selected fields have been modified.

- **Nonrepudiation:** this service guarantees that an entity once involved in a communication cannot later deny its involvement. This service may take one or both of two forms:

  - *Nonrepudiation with proof of origin:* the recipient of the data is provided with proof of the origin of the data. This will protect against any attempt by the sender to falsely deny sending the data or their contents. A digital signature is an example of providing nonrepudiation with proof of origin (cf. p.10, X.800).
  - *Nonrepudiation with proof of delivery:* the sender of data is provided with proof of delivery of the data. This will protect against any subsequent attempt by the recipient to falsely deny receiving the data or their contents (cf. p.10, X.800).

15.1.3 Security Mechanisms

A security mechanism is a process (or a device incorporating such a process) that can be used in a system to implement a security service that is provided by or within the system. Some examples of security mechanisms are authentication exchange, checksums, digital signatures, encryption and traffic padding (cf. p.153, RFC 2828). Security mechanisms are divided into two groups: *specific security mechanisms*, which may be incorporated in a specific protocol layer, and *pervasive security mechanisms*, which are not specific to any particular protocol layer. The concepts below are taken from the X.800 Recommendations

- Specific security mechanisms

  - *Encipherment:* encipherment can provide confidentiality of either data or traffic flow information by converting the original information into a form that is not intelligible. Encipherment algorithms may be reversible or irreversible. Two general classifications of reversible encipherment algorithm are *symmetric* (i.e. secret key) and *asymmetric* (i.e. public key).
  - *Digital signature:* this mechanism attaches some special information to the transmitted data, enabling the recipient to verify the source as well as the integrity of the data. The term digital signature goes hand-in-hand with public key cryptography.
  - *Access control:* this mechanism may use the authenticated identity of an entity, information about the entity or capabilities of the entity to grant access rights to the entity.
  - *Data integrity:* two aspects of data integrity are: (i) the integrity of a single data unit or field; (ii) the integrity of a stream of data units or fields. In general, different mechanisms are used to provide these two types of integrity service, although provision of the second without the first is not practical.
  - *Authentication exchange:* peer entity authentication is assisted by means of information exchange.
  - *Traffic padding mechanism:* traffic padding mechanisms can be used to provide various levels of protection against traffic analysis. Traffic padding is done by inserting bits into gaps in data streams.
  - *Routing control:* this mechanism allows a proper choice of routes for transferring information. End systems may wish to instruct the network service provider to establish a connection via a different route for a more secure communication.
  - *Notarization mechanism:* this mechanism needs the involvement of a third party to ensure certain properties of data exchange between the two entities.

- Pervasive security mechanisms

  - *Trusted functionality:* may be used to extend the scope, or to establish the effectiveness, of other security mechanisms. Any functionality which provides access to security mechanisms should be trustworthy.
  - *Security labels:* resources including data items may have security labels associated with them, e.g. to indicate a sensitivity level. It is often necessary to convey the appropriate security label with data in transit.
  - *Event detection:* security-relevant event detection includes the detection of apparent violations of security and may also include detection of 'normal' events.

- *Security audit trails:* provide a valuable security mechanism, as potentially they permit detection and investigation of breaches of security by permitting a subsequent security audit. A security audit is an independent review and examination of system records and activities in order to test for adequacy of system controls, to ensure compliance with established policy and operational procedures, to aid in damage assessment and to recommend any indicated changes in controls, policy and procedures.
- *Security recovery:* security recovery deals with requests from mechanisms such as event handling and management functions and takes recovery actions as the result of applying a set of rules. These recovery actions may be of three kinds: immediate, temporary or long term.

15.1.4 Relationships between Security Services and Mechanisms

Table 15.1 illustrates the relationships between security services and security mechanisms.

Authentication services need a digital signature mechanism since it helps to authenticate the sender (from both directions). An encipherment mechanism is also needed here. In addition, peer entity authentication also needs authentication exchange to authenticate both sides during the connection time.

Nonrepudiation services need digital signatures as proof of the participants' identities. However, these services do not need encipherment since the exchanged data may not need to be confidential (e.g. normal emails). These services, instead, need a data integrity mechanism to guarantee that the received information is unchanged and a notarization mechanism to convince the other side of one side's authenticity and to solve any possible disputes. Similar explanations come with other combinations of services and mechanisms.

15.1.5 Placements of Security Services and Mechanisms

Tables 15.2 and 15.3 outline the placement of security services and mechanisms in OSI protocol layers.

15.2 Wired Equivalent Privacy (WEP)

As a part of the IEEE 802.11 wireless network standard, *wired equivalent privacy* (WEP) is used to secure IEEE 802.11 wireless networks (Wi-Fi). Since Wi-Fi technology uses radio communication, which is vulnerable to eavesdropping, it is necessary to have mechanisms to guarantee privacy in Wi-Fi networks. Once designed, WEP is supposed to provide comparable security to traditional wired networks. Its main purpose is to protect Wi-Fi communications from eavesdropping. In addition, WEP aims to prevent unauthorized access to Wi-Fi networks.

15.2.1 How Does WEP Work?

WEP secures communications between mobile clients and wireless access points by encryption using shared keys. To access a Wi-Fi network, a mobile client needs to have a WEP key that matches the one configured at its appropriate access point. Having checked this WEP key for validity, the access point grants the mobile client access to the network. Figure 15.1 illustrates a simple WEP-based WLAN configuration.

Table 15.1 Relationship between security services and security mechanisms (cf. p.15, Recommendation X.800)

	Encipherment	Digital signature	Access control	Data integrity	Authentication exchange	Traffic padding	Routing control	Notarization
Data origin authentication	Y	Y	—	—	—	—	—	—
Peer entity authentication	Y	Y	—	—	Y	—	—	—
Access control	—	—	Y	—	—	—	—	—
Connection confidentiality	Y	—	—	—	—	—	Y	—
Connectionless confidentiality	Y	—	—	—	—	—	Y	—
Selective field confidentiality	Y	—	—	—	—	—	—	—
Traffic flow confidentiality	Y	—	—	—	—	Y	Y	—
Connection integrity with recovery	Y	—	—	Y	—	—	—	—
Connection integrity without recovery	Y	—	—	Y	—	—	—	—
Selective field connection integrity	Y	—	—	Y	—	—	—	—
Connectionless integrity	Y	Y	—	Y	—	—	—	—

Selective field connectionless integrity	Y	Y	—	Y	—
Nonrepudiation with proof of origin	—	Y	—	Y	Y
Nonrepudiation with proof of delivery	—	Y	—	Y	Y

Notes: Y: the mechanism is considered to be appropriate, either on its own or in combination with other mechanisms
—: the mechanism is considered not to be appropriate

Table 15.2   Placement of security services in OSI protocol layers (cf. p.27, Recommendation X.800)

Service	Layers						
	1	2	3	4	5	6	7*
Data origin authentication	—	—	Y	Y	—	—	Y
Peer entity authentication	—	—	Y	Y	—	—	Y
Access control	—	—	Y	Y	—	—	Y
Connection confidentiality	Y	Y	Y	Y	—	Y	Y
Connectionless confidentiality	—	Y	Y	Y	—	Y	Y
Selective field confidentiality	—	—	—	—	—	Y	Y
Traffic flow confidentiality	Y	—	Y	—	—	—	Y
Connection integrity with recovery	—	—	—	Y	—	—	Y
Connection integrity without recovery	—	—	Y	Y	—	—	Y
Selective field connection integrity	—	—	—	—	—	—	Y
Connectionless integrity	—	—	Y	Y	—	—	Y
Selective field connectionless integrity	—	—	—	—	—	—	Y
Nonrepudiation with proof of origin	—	—	—	—	—	—	Y
Nonrepudiation with proof of delivery	—	—	—	—	—	—	Y

Y: service is provided within the layer mentioned.
—: service is not provided within the layer mentioned
* It should be noted, with respect to layer 7, that the application process may, itself, provide security services

Table 15.3   Placement of security mechanisms in OSI protocol layers

Mechanism	Layers						
	1	2	3	4	5	6	7
Encipherment	Y	Y	Y	Y	—	Y	—
Digital signature	—	—	Y	Y	—	Y	Y
Access control	—	—	Y	Y	—	—	Y
Data integrity	—	—	Y	Y	—	Y	—
Authentication exchange	—	—	Y	Y	—	—	Y
Traffic padding	—	—	Y	—	—	—	Y
Routing control	—	—	Y	—	—	—	—
Notarization	—	—	—	—	—	Y	Y

Y: mechanism is provided within the layer mentioned
—: mechanism is not provided within the layer mentioned

As it uses the same WEP key for all clients of an access point, WEP's sole purpose is to prevent information eavesdropping from outside attackers. It does not offer protection against all other clients on the network. For confidentiality purposes, WEP encryption/decryption uses the RC4 stream cipher. To form a 64-bit RC4 key, WEP uses a 24-bit initialization vector (IV) and a 40-bit key. The concatenation of these two fields creates a 64-bit WEP key. In the extended configuration with 128-bit WEP key, the key size is 104 bits plus 24 bits for IV.

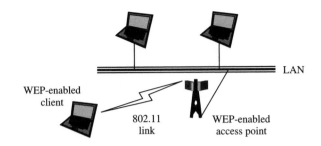

Figure 15.1   Simple WEP-based WLAN configuration

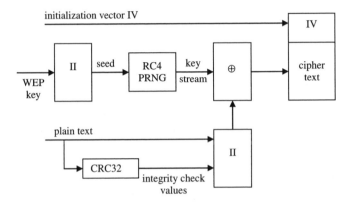

Figure 15.2   WEP encryption principles (cf. Figure 43a, p.37, 802.11i-2004 standard)

Figure 15.2 represents WEP encryption, which works as follows. On the sending side, the sender uses the initialization vector and 40-bit WEP key to create a key stream. The key stream is then XORed with the plain text to produce the cipher text. A cyclic redundancy check, CRC-32, is also derived from the plain text to check for message integrity at the receiving side.

To decrypt the cipher text, the receiver uses the basic principle of the XOR operator, i.e.

$$ciphertext = keystream \oplus plaintext$$

$$keystream \oplus ciphertext = keystream \oplus (keystream \oplus plaintext)$$

$$= (keystream \oplus keystream) \oplus plaintext$$

$$= plaintext$$

Thus, on the receiving side, the receiver, who has the same WEP key, uses the IV in the receiving message and recovers the same key stream as on the sending side. The receiver then XORs the received cipher text with the key stream to recover the plain text, following the rule above. The WEP decryption principle is represented in Figure 15.3.

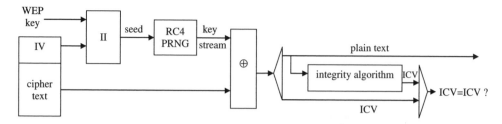

Figure 15.3 WEP decryption principles (cf. Figure 43b, p.38, 802.11i-2004 standard)

The CRC-32 created at the sending side is used as the integrity check value (ICV) to verify if the received message has been altered during transmission. Only when this value matches with that created by the integrity algorithm is the message considered valid.

15.2.2 WEP Weaknesses

Pieces of work and presentations published by researchers have shown that WEP has security flaws. According to research published by the ISAAC group (http://www.isaac.cs.berkeley .edu/) at the University of California at Berkeley, WEP's operation using stream ciphers is vulnerable to several kinds of attack.

If the cipher text is attacked on its way to the receiver and some bits are changed, the corresponding bits in the decrypted plain text will also be changed. Also, if an attacker can capture two cipher texts encrypted using the same key stream and XOR these two cipher texts, then this result is also the XOR of the two appropriate plain texts. Having gained and collected this information, the attacker can use statistical attacks to recover the plain texts. The more frequently a key stream is used for encryption and is captured by an attacker, the more easily he can perform statistical attacks. When the attacker recovers one of the plain texts, he can also recover the others.

The designers of WEP have acknowledged these flaws and have created counter measures to increase its security. To deal with the integrity of packets in transit, they use the integrity check field (IC) in the packet's header. To avoid using the same key stream to encrypt packets, they use the initialization vector (IV) field to produce different encryption keys (Figure 15.4). However, these proposals do not solve WEP's security problems completely.

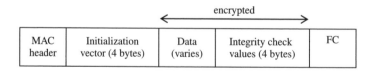

Figure 15.4 WEP packet format

The integrity check field referred to above is a checksum with 32-bit CRC and is also encrypted. However, research results show that since CRC-32 is linear, it is possible to compute the bit difference of two CRCs based on the bit difference of the messages over which they are taken, meaning that, if attackers change a bit in the original message, they can some-how calculate changes to other bits in the CRC field so that the checksum is still correct. With this technique, the effort of WEP's creator to guarantee message integrity becomes obsolete.

Another effort to increase WEP security is the use of a 24-bit initialization vector to pro-duce a different key stream. However, research results also show that, using only 24 bits to differentiate the key streams is not enough since it can only produce around 16 million keys. In a heavy-load network, transporting 16 million packets only takes a short period of time. This means that there will soon be two packets encrypted with the same key stream. It also leads to statistical attack, as mentioned above.

The following are several kinds of attack reported by the ISAAC group at the University of California at Berkeley (http://www.isaac.cs.berkeley.edu/).

15.2.2.1 Passive Attack to Decrypt Traffic

In this type of attack, a passive eavesdropper can intercept all wireless traffic until an IV collision occurs. By XORing two packets that use the same IV, the attacker obtains the XOR of the two plain text messages. The resulting XOR can be used to infer data about the contents of the two messages.

Since IP traffic is often very predictable and includes a lot of redundancy, the attacker can eliminate many possibilities for the contents of messages. Further educated guesses about the contents of one or both of the messages can be used to statistically reduce the space of possible messages, and in some cases it is possible to determine the exact contents.

When such statistical analysis is inconclusive based on only two messages, the attacker can look for more collisions of the same IV. With only a small factor in the amount of time necessary, it is possible to recover a modest number of messages encrypted with the same key stream, and the success rate of statistical analysis grows quickly. Once it is possible to recover the entire plain text for one of the messages, the plain text for all other messages with the same IV follows directly, since all the pairwise XORs are known.

An extension to this attack uses a host somewhere on the Internet to send traffic from the outside to a host on the wireless network. The contents of such traffic will be known to the attacker, yielding known plain text. When the attacker intercepts the encrypted version of his message sent over 802.11, he will be able to decrypt all packets that use the same initialization vector.

15.2.2.2 Active Attack to Inject Traffic

Let's assume that an attacker knows the exact plain text for an encrypted message. Based on the basic property that $RC4 (x) \oplus x \oplus y = RC4 (y)$, where x and y are two separate bit streams, the attacker may use this knowledge to encrypt other packets. The procedure involves constructing a new message, calculating the CRC-32 and performing bit flips on the original encrypted message to change the plain text to the new message. This packet can now be sent to the access point or mobile station and will be accepted as a valid packet.

A slight modification to this attack makes it much more dangerous. Even without complete knowledge of the packet, it is possible to flip selected bits in a message and successfully adjust the encrypted CRC to obtain a correctly encrypted version of a modified packet. If the attacker has partial knowledge of the contents of a packet, he can intercept it and perform selective modification to it. For example, it is possible to alter commands that are sent to the shell over a telnet session or interactions with a file server.

15.2.2.3 Active Attack from Both Ends

The previous attack can be extended further to decrypt arbitrary traffic. In this case, the attacker makes a guess not about the contents, but rather the headers of a packet. This information is usually quite easy to obtain or guess; in particular, all that is necessary to guess is the destination IP address. Armed with this knowledge, the attacker can flip appropriate bits to transform the destination IP address to send the packet to a machine he controls, and transmit it using a rogue mobile station. Most wireless installations have Internet connectivity; the packet will be decrypted successfully by the access point and forwarded unencrypted through appropriate gateways and routers to the attacker's machine, revealing the plain text. If a guess can be made about the TCP headers of the packet, it may even be possible to change the destination port on the packet to port 80, which will allow it to be forwarded through most firewalls.

15.2.2.4 Table-based Attack

The space for initialization vectors allows an attacker to build a decryption table. Once he learns the plain text for some packet, he can compute the RC4 key stream generated by the IV used. This key stream can be used to decrypt all other packets that use the same IV. Over time, perhaps using the techniques above, the attacker can build up a table of IVs and corresponding key streams. This table requires a fairly small amount of storage ($\sim$15 GB); once it is built, the attacker can decrypt every packet that is sent over the wireless link.

15.2.2.5 Monitoring

Despite the difficulty of decoding a 2.4 GHz digital signal, hardware to listen to 802.11 transmissions is readily available to attackers in the form of consumer 802.11 products. The products possess all the necessary monitoring capabilities and all that remains for attackers is to convince it to work for them. Although 802.11 equipment is generally designed to disregard encrypted content for which it does not have the key, attackers have been able to intercept WEP-encrypted transmissions successfully by changing the configuration of the device drivers. They were able to confuse the firmware enough that the cipher text of unrecognized packets was returned to them for further examination and analysis.

Active attacks (those requiring transmission, not just monitoring) appear to be more difficult, yet not impossible. Many 802.11 products come with programmable firmware, which can be reverse engineered and modified to provide the ability to inject traffic to attackers. Granted, such reverse engineering is a significant time investment, but it's important to note that it's a one-time cost. A competent group of people can invest this effort and then distribute the rogue firmware through underground circles, or sell it to parties interested in corporate espionage. The latter is a highly profitable business, so the time investment is easily recovered.

15.3 Wi-Fi Protected Access (WPA)

Wi-Fi Protected Access (WPA) has been proposed by the Wi-Fi Alliance as a replacement for WEP in enhancing security in Wi-Fi networks, while waiting for the IEEE 802.11i standard to be developed and approved. WPA is now a subset of the IEEE 802.11i standard. WPA can work under two modes: WPA Enterprise, which uses 802.1X for authentication, and WPA Personal, which uses a preshared key for authentication. WPA also uses an algorithm called the *temporal key integrity protocol* (TKIP) for encryption and another algorithm called *Michael* for message integrity.

Like WEP, data encryption in WPA uses an RC4 stream cipher, with a key of 128 bits and an initialization vector of 48 bits. However, WPA uses TKIP to change the keys dynamically. The ability to change key plus the bigger space for the IV makes a key recovery attack in WPA theoretically impossible. For message integrity, WPA uses message integrity code (MIC) instead of WEP's CRC, which helps to prevent replay attacks.

By increasing the size of the encryption keys and initialization vector space, WPA reduces the number of packets sent with related keys. In addition, by creating a better message integrity check, WPA makes breaking into a Wi-Fi network much more difficult.

15.3.1 How Does WPA Work?

WPA can work under two modes: *WPA Enterprise* and *WPA Personal*. In the WPA Enterprise mode, there are mobile clients, an access point and an authentication server (RADIUS or LDAP). In this model, when a mobile client wants to access a network, it contacts the appropriate access point for authentication. The client sends authentication information to the access point. This information is then forwarded to the authentication server for validation. After checking for valid credentials, the authentication server instructs the access point to allow the client to access the network. The server also sends an encryption key to the access point and the client. The client uses this key to encrypt information it exchanges with the access point. Figure 15.5 shows the configuration of the WPA Enterprise mode. The WPA Enterprise mode is normally used in companies and organizations with large numbers of users and available information technology infrastructure.

The WPA Personal mode, meanwhile, is used for home users or small offices where the deployment of an authentication server is not feasible. Thus, WPA Personal is sometimes referred to as *WPA SOHO* (Small Office, Home Office). It is also called the *preshared key (PSK) mode*. In this PSK mode, a password is manually entered at the access point as well as being given to the mobile client. When a client wants to join the network, the access point first checks if the password of the client matches its own. If so, the access point grants the client access to the network. It then delivers the encryption key to the mobile client and the information exchange starts. WPA Personal mode is represented in Figure 15.6.

WPA can be summarized as: WPA = 802.1x + EAP + TKIP + MIC (for WPA Personal, EAP is replaced by PSK). Fields in a WPA packet are shown in Figure 15.7. The security of WPA is based on secured authentication, strong encryption and message integrity. In the following section, we present each in turn.

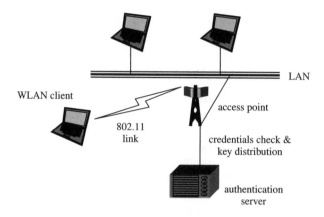

Figure 15.5   WPA Enterprise mode

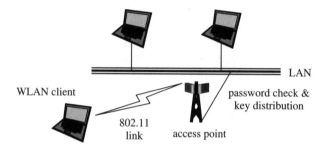

Figure 15.6   WPA Personal mode

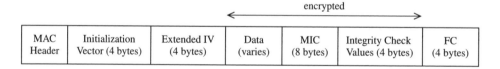

Figure 15.7   WPA packet format

15.3.1.1 Authentication

WPA authentication conforms to the IEEE 802.1x specification, which uses the extensible authentication protocol (EAP) for centralized user and/or wireless network authentication, as well as encryption key management and distribution. WPA supports two modes of authentication for different purposes, as indicated in Table 15.4.

In the Enterprise mode, WPA uses 802.1x and EAP and requires an authentication server. WPA uses EAP to transport authentication information for mobile clients and wireless networks. EAP is an extensible authentication protocol and supports different authentication methods, such as passwords, digital certificates and smartcards. Mobile clients can use any of the supported authentication methods installed in the authentication server. If all the authentication methods fail, the access point rejects the credentials of a mobile client.

Table 15.4 WPA authentication support

Authentication mode	Protocol	Authentication server needed
WPA Enterprise	802.1x + EAP	Yes
WPA SOHO	802.1x + preshared key (PSK)	No

An access point with WPA support broadcasts beacon messages. When a mobile terminal is close to that access point, it associates its SSID with the access point. If association is successful, the mobile terminal starts the authentication process.

Having authenticated, two sets of keys, namely pairwise keys and groupwise keys, are delivered to the mobile client. The keys are used for encrypting 802.11 packets before being sent over the air. The groupwise keys are used for all mobile clients connected to the same access point. The pairwise keys are unique between a mobile client and the access point. This 802.1x key distribution mechanism solves the WEP authentication problem where all mobile clients share the same WEP key for authentication and data encryption.

In Personal mode, since there is no authentication server, a simple password-matching scheme is used. This authentication mode is called *preshared key* (WPA-PSK). In this mode, a single preshared key (or password) is manually entered at each mobile client and at the access point. The mobile client is granted access to the network if its password matches the one held by the access point.

Regardless of the WPA authentication method in use, after successful authentication, TKIP is used for encrypting transmitted messages. This encryption method makes WPA-PSK different from WEP. This is also the reason why WPA Personal mode is more secure than WEP.

15.3.1.2 Encryption

The problem with WEP's encryption is mainly due to the IV sent over the air without encryption. In a heavy-load Wi-Fi network, the IV repeats once every few hours. By capturing packets with the same value of IV, attackers can figure out the WEP key by repeatedly XORing the cipher text, and then can illegally access the network.

WPA, with the use of the TKIP, solves this problem of WEP by:

- using a longer IV (48 bits);
- increasing the key size from 40 to 128 bits;
- renewing the encryption key every 10 000 packets;
- using per packet key mixing of the IV.

The encryption key is produced by appending a 48-bit IV, a 104-bit RC4 key and the client's MAC address, then finally putting the result through a mixing function to create a 128-bit RC4 encryption key value.

15.3.1.3 Message Integrity

The message integrity check (MIC) in WPA, as implied by its name, provides the integrity function. As mentioned above, the CRC-32 of WEP does not provide message integrity and an MIC one-way hash function is used to replace the CRC-32 checksum used in WEP. In the MIC scheme, the receiver and transmitter each compute, and then compare, the MIC value. If they do not match, the data are assumed to have been changed in transit and the packet is dropped.

MIC uses an algorithm, called Michael, to compute a 64-bit value by using the source and destination MAC addresses and the data field. By computing MIC over source/destination addresses, the packet data are associated with the sender and receiver, preventing attacks based on packet forgery.

15.3.2 WEP and WPA Comparison

Table 15.5 presents a brief comparison between WEP and WPA.

Table 15.5 Comparison between WEP and WPA (Reproduced, with permission from Wi-Fi Alliance, 2003)

	WEP	WPA
	Flawed, cracked by scientists and hackers	Fixes all WEP flaws
	40-bit keys	128-bit keys
Encryption	Static key – same key used by everyone on the network	Dynamic session keys, i.e. per user, per session, per packet keys
	Manual distribution of keys – hand typed into each device	Automatic distribution of keys
Authentication	Flawed, used WEP key itself for authentication	Strong user authentication, utilizing 802.1X and EAP

In summary, if an organization needs stronger security for its wireless network, it is strongly recommended that it replaces the use of WEP by WPA.

15.3.3 WPA2

In 2004, the IEEE approved the full IEEE 802.11i specification, which was quickly followed by a new interoperability testing certification from the Wi-Fi Alliance known as *WPA2*. WPA2 is based on the robust security network (RSN) mechanism and provides support for all of the mechanisms available in WPA.

WPA and WPA2, in fact, share many similarities. They both use EAP and 802.1x for authentication. They also work in Enterprise and Personal modes. The sole difference between them is the encryption mechanism. While WPA uses TKIP with an RC4 stream cipher, WPA2 uses the advanced encryption standard (AES) with a CCMP (counter mode with cipher block chaining message authentication code protocol) encryption mechanism. AES is stronger than RC4, which may be required by some corporate and government users.

As of March 2006, the WPA2 certification became mandatory for all new equipment certified by the Wi-Fi Alliance, ensuring that any state-of-the-art hardware will support both WPA and WPA2.

References

Akyildiz, I. F., Su, W., Sankarasubramaniam, Y. and Cayirci, E. (2002) 'Wireless Sensor Networks: A Survey', *Computer Networks*, **38**, 393–422.

Akyildiz, I. F., Wang, X. and Wang, W. (2005) 'Wireless mesh networks: a survey,' *Computer Networks*, **47**, 445–487.

Anderson, R. and Kuhn, M. (1997) 'Low Cost Attacks on Tamper Resistant Devices', proceedings of the 1997 Security Protocols Workshop, Paris, Springer *LNCS*, **1361**, 125–136.

Anderson, R., Chan, H. and Perrig, A. (2004) 'Key infection: Smart trust for smart dust,' Proceedings of the 12th IEEE International Conference on Network Protocols, ICNP'04, pp. 206–215.

ANSI/IEEE (1999) Standard 802.11/ Standard 802.11a/ Standard 802.11b/ Standard 802.11i.

Arisha, K., Youssef, M. and Younis, M. (2002) 'Energy-Aware TDMA-Based MAC for Sensor Networks,' in *Proceedings of the IEEE Workshop on Integrated Management of Power Aware Communications, Computing and Networking (IMPACCT 2002)*, New York City, New York, May 2002.

Asokan, N. and Ginzboorg, P. (2000) 'Key agreement in ad-hoc networks,' *Computer Communications*, **23** (17), 1627–1637.

Atallah, M., Bryant, E. and Stytz, M. (2004) 'A Survey of Anti-Tamper Technologies', *CrossTalk: The Journal of Defense Software Engineering*, November.

Aune, F. (2004) 'Cross Layer Design Tutorial', Norwegian University of Science and Technology, Trondheim, November.

Balenson, D., McGrew, D. and Sherman, A. (2000) 'Key Management for Large Dynamic Groups: One-Way Function Trees and Amortized Initialization,' IETF Internet-Draft, August 25 2000, draft-irtf-smug-groupkeymgmt-oft-00.txt.

Balfanz, D., Smetters, D., Stewart, P. and Wong, H. (2002) 'Talking to strangers: Authentication in ad hoc wireless networks,' in Symposium on Network and Distributed Systems Security (NDSS '02), San Diego, California, USA.

Barakat, C., Altman, E. and Dabbous, W. (2000) 'On TCP performance in a heterogeneous network: a survey,' *IEEE Communications Magazine*, **38**(1), 40–46.

Basagni, S., Herrin, K., Bruschi, D. and Rosti, E. (2001) 'Secure pebblenets,' MobiHoc 2001, pp. 156–163.

Becker, K. and Wille, U. (1998) 'Communication complexity of group key distribution,' *proceedings of the 5th ACM conference on Computer and Communication Security*, pp. 1–6.

Bluetooth SIG (2004) 'Bluetooth Core Specification Version 2.0 + Enhanced Data Rate,' Bluetooth SIG.

Boneh, D. and Boyen, X. (2004) 'Secure Identity Based Encryption Without Random Oracles,' proceedings of Crypto 2004, *LNCS*.

Boneh, D. and Franklin, M. (2001) 'Identity-based encryption from the weil pairing,' *proceedings of Crypto 2001*, pp. 213–229.

Boneh, D., Boyen, X. and Goh, E. J. (2005) 'Hierarchical Identity Based Encryption with Constant Size Ciphertext,' proceedings of EUROCRYPT'05, *LNCS*, **3494**, 440–456.

Boyen, X. (2003) 'Multipurpose Identity-Based Signcryption – A Swiss Army Knife for Identity-Based Cryptography,' proceedings of Crypto 2003, *LNCS*, **2729**, 383–399.

Buchegger, S. and Le Boudec, J. (2002) 'Performance Analysis of the CONFIDANT Protocol (Cooperation of Nodes: Fairness in Dynamic Ad Hoc Networks,' MobiHoc 2002.

Bulusu, N., Heideman, J. and Estrin, D. (2000) 'GPS-less low cost outdoor localization for very small devices,' *IEEE Personal Communication*, **7**, 28–34.

Burmester, M. and Desmedt, Y. (1994) 'A secure and efficient conference key distribution system,' *proceedings of EUROCRYPT' 94*, pp. 275–286.

Cagalj, M., Capkun, S. and Hubaux, J. P. (2006) 'Key Agreement in Peer-to-Peer Wireless Networks,' *Proceedings of the IEEE*, **94**(.2), 467–478.

Camtepe, S. A. and Yener, B. (2005) '*Key Distribution Mechanisms for Wireless Sensor Networks: a Survey*,' Technical report TR-05-07, Rensselaer Polytechnic Institute, NY, USA.

Canetti, R., Garay, J., Itkis, G., Micciancio, D. and Naor, M. (1999) 'Multicast Security: A Taxonomy and Efficient Constructions,' *proceedings of INFOCOMM'99*.

Capkun, S., Buttyán, L. and Hubaux, J. P. (2003a) 'Self-Organized Public-Key Management for Mobile Ad Hoc Networks,' *IEEE Transactions on Mobile Computing*, **2**(1), 1–13.

Capkun, S. and Hubaux, J. (2003) 'BISS: Building secure routing out of an incomplete set of security associations,' WiSE.

Capkun, S., Hubaux, J. P. and Buttyán, L. (2003b) 'Mobility Helps Security in Ad Hoc Networks,' *proceedings of MobiHoc'03*.

Capkun, S., Hubaux, J. P. and Buttyán, L. (2006) 'Mobility Helps Peer-to-Peer Security,' *IEEE Transactions on Mobile Computing*, **5**(1), 43–51.

Cardei, M. and Wu, J. (2005) 'Coverage in Wireless Sensor Networks,' in *Handbook of Sensor Networks: Compact Wireless and Wired Sensing Systems*, CRC Press.

Carter, S. and Yasinsac, A. (2002) 'Secure position aided ad hoc routing protocol,' IASTED International Conference on Communications and Computer Networks (CCN02).

Cayirci, E. (2003) 'Data Aggregation and Dilution by Using Modulus Addressing in Wireless Sensor Networks', *IEEE Communications Letters*, August.

Cayirci, E. and Akyildiz, I.F. (2002) 'User Mobility Pattern Scheme for Location Update and Paging in Wireless Systems', *IEEE Transactions on Mobile Computing*, **1**, 236–247.

Cayirci, E. and Akyildiz, I.F. (2003) 'Optimal Location Area Design to Minimize Registration Signaling Traffic in Wireless Systems', *IEEE Transactions on Mobile Computing*, **2**, 76–85.

Cayirci, E. and Coplu, T. (in press) 'SENDROM: Sensor Networks for Disaster Relief Operations Management,' *ACM/Kluwer Wireless Networks*.

Cayirci, E. and Ersoy, C. (2002) 'Application of 3G PCS Technologies to the Rapidly Deployable Networks', *IEEE Network Magazine*, September/October, pp. 20–27.

Cayirci, E. and Nar, P. (2005) 'Power Controlled Sensor MAC Protocol,' EWSN'2005.

Cayirci, E., Coplu, T. and Emiroglu, O. (2005) 'Power Aware Many To Many Data Centric Routing In Wireless Sensor and Actuator Networks,' EWSN'2005.

Cayirci, E., Cimen, C. and Coskun, V. (2006a) 'Querying Sensor Networks by Using Dynamic Task Sets,' *Computer Networks*, **50**(7), 938–952.

Cayirci, E., Tezcan, H. and Coskun, V. (2006b) 'Wireless Sensor Networks for Underwater Surveillance Systems,' *AdHoc and Sensor Networks*, **4**(4), 431–446.

Certicom Corp. (2004) 'MQV: Efficient and Authenticated Key Agreement,' *Code & Cipher*, Certicom's bulletin of security and cryptography, Crypto Column, **1**(2).

Cha, J. C. and Cheon, J. H. (2002) 'An Identity-Based Signature from Gap Diffie–Hellman Groups,' *Cryptology eprint Archive*, Report 2002/18.

Chan, H. and Perrig, A. (2003) 'Security and Privacy in Sensor Networks' in *IEEE Computer Magazine*, October.

Chan, H., Perrig, A. and Song, D. (2003) 'Random key predistribution schemes for sensor networks,' proceedings of the IEEE Symposium on Security and Privacy, IEEE Computer Society, pp. 197–213.

Chang, H. and Atallah, M. (2001) 'Protecting Software Code By Guards,' proceedings of ACM Workshop on Security and Privacy in Digital Rights Management, Philadelphia, PA, pp.160–175.

Chen, L. and Kudla, C. (2003) '*Identity Based Authenticated Key Agreement Protocols from Pairings*,' HP Technical Report HPL-2003-25.

Collberg, C. and Thomborson, C. (2002) 'Watermarking, Tamper-Proofing, and Obfuscation Tools for Software Protection,' *IEEE Transactions on Software Engineering*, **28**(8), 735–746.

Collberg, C., Thomborson, C. and Low, D. (1997) '*A Taxonomy of Obfuscating Transformations*,' Department of Computer Science, University of Auckland, New Zealand.

Coskun, V., Cayirci, E., Levi, A. and Sancak, S. (2006) 'Quarantine Region Scheme to Prevent Spam Attacks in Wireless Sensor Networks,' *IEEE Transactions on Mobile Computing*, 5(8), 1074–1086.

Dam, T. and Langendoen, K. (2003) 'An Adaptive Energy-Efficient MAC Protocol for Wireless Sensor Networks,' ACM SenSys, November.

Deng, J., Han, R. and Mishra, S. (2003) 'INSENS: intrusion tolerant routing in wireless sensor networks,' 23rd IEEE International Conference on Distributed Computing Systems (ICDCS).

Deng, J., Han, R. and Mishra, S. (2004) '*Countermeasures against traffic analysis in wireless sensor networks,*' Technical Report CU-CS-987-04, University of Colorado at Boulder.

Deng, J., Han, R. and Mishra, S. (2005) 'Countermeasures Against Traffic Analysis Attacks in Wireless Sensor Networks', in first IEEE/CerateNet Conference on Security and Privacy in Communication Networks (SecureComm 2005), Athens, Greece, September 2005, pp. 113–124.

Desmedt, Y. G. (1994) 'Threshold Cryptography,' *European Transactions on Telecommunications*, 5(4), 449–457.

Diffie, W. and Hellman, M. E. (1976) 'New Directions in Cryptography,' *IEEE Transactions on Information Theory*, **IT-22**(6), 644–654.

Ding, M., Chen, D., Xing, K. and Cheng, X. (2005) 'Localized Fault Tolerant Event Boundary Detection in Sensor Networks', INFOCOM 2005.

Di Pietro, R., Mancini, L. V., Law, Y. W., Etalle, S. and Havinga, P. (2003) 'LKHW: A directed diffusion based secure multicasting scheme for wireless sensor networks,' First International Workshop on Wireless Security and Privacy (WiSPr'03).

Djenouri, D., Khelladi, L. and Badache, N. (2005) 'A Survey of Security Issues in Mobile Ad Hoc and Sensor Networks,' *IEEE Communications Surveys & Tutorials*, 7(4), fourth quarter 2005.

Doherty, L., Pister, K. S. J. and Ghaoui, L. E. (2001) 'Convex position estimation in wireless sensor networks,' Infocom'01, Anchorage.

Douceur, J. R. (2002) 'The Sybil Attack,' 1st International Workshop on Peer-to-Peer Systems (IPTPS'02), pp. 251–260.

Du, W., Deng, J., Han, Y. S. and Varshney, P. (2003) 'A Pairwise Key Pre-distribution Scheme for Wireless Sensor Networks,' proceedings of CCS'03.

Du, W., Deng, J., Han, Y. S., Chen, S. and Varshney, P. K. (2004) 'A Key Management Scheme for Wireless Sensor Networks Using Deployment Knowledge,' proceedings of INFOCOM '04.

ECMA (2005) International Standards 368 and 369, High Rate Ultra Wideband PHY and MAC Standards.

Elson, J., Girod, L. and Estrin, D. (2002) *Fine grained network time synchronization using reference broadcasts*, OSDI, Boston.

Erdogan, A., Cayirci, E. and Coskun, V. (2003) 'Sectoral Sweepers for Task Dissemination and Location Estimation in Ad Hoc Sensor Networks', MilCom'2003, Boston.

Eschenauer, L. and Gligor, V. D. (2002) 'A key-management scheme for distributed sensor networks,' proceedings of the 9th Conference on Computer Communication Security (CCS2002), pp. 41–47.

Fiat, A. and Shamir, A. (1987) 'How to Prove Yourself: Practical Solutions to Identification and Signature Problems,' proceedings of Crypto '86, pp.186–194.

Fokine, K. (2002) '*Key Management in Ad Hoc Networks*,' Masters thesis, LiTH-ISY-EX-3322-2002, Lindköpings tekniska högskola.

Fonseca, E. and Festag, A. (2006) '*A Survey of Existing Approaches for Secure Ad Hoc Routing and Their Applicability to VANETS*,' NEC Technical Report NLE-PR-2006-19.

Ganeriwal, S., Kumar, R. and Srivastava, M. D. (2003) 'Timing Synch Protocol for Sensor Networks,' first ACM Conference on Embedded Networked Sensor Systems (SenSys), November 2003, pp 139–149.

Ganeriwal, S., Capcun, S., Han, C. and Srivastava, M. B. (2005) 'Secure Time Synchronization Service for Sensor Networks,' WiSE.

Gennaro, R., Jarecki, S., Krawczyk, H. and Rabin, T. (1999) 'Secure Distributed Key Generation for Discrete-Log Based Cryptosystems,' proceedings of EUROCRYPT'99.

Golle, P., Greene, D. and Staddon, J. (2004) 'Detecting and correcting malicious data in VANETs,' first ACM Workshop on Vehicular Ad Hoc Networks (VANET).

Gruteser, M., Schelle, G., Jain, A., Han, R. and Grunwald, D. (2003) "Privacy-aware location sensor networks', in 9th USENIX Workshop on Hot Topics in Operating Systems (HotOS IX).

Haas, Z. J. and Liang, B. (1999) 'Ad Hoc Mobility Management with Uniform Quorum Systems,' *IEEE/ACM Transactions on Networking*, **7**(2), 228–240.

Hammes, T. X. (2007) 'Fourth Generation Warfare Evolves, Fifth Emerges,' *Military Review*, May–June, pp. 14–23.

Havinga, P. and Smit, G. (2000) 'Energy-efficient TDMA medium access control protocol scheduling,' *proceedings of the Asian International Mobile Computing Conference (AMOC 2000)*, November 2000.

Hegland, A. M., Winjum, E., Mjølsnes, S. F., Rong, C., Kure, Ø. and Spilling, P. (2006) 'A Survey of Key Management in ad hoc Networks,' *IEEE Communications Surveys & Tutorials*, **8**(3), 48–66.

Heinzelman, W. R., Kulik, J. and Balakrishan, H. (1999) 'Adaptive Protocols for Information Dissemination in Wireless Sensor Networks,' MobiCom'99, pp. 174–185.

Heinzelman, W. R., Chandrakasan, A. and Balakrishnan, H. (2000) 'Energy-Efficient Communication Protocol for Wireless Microsensor Networks,' IEEE Hawaii International Conference on System Sciences, pp. 1–10.

Helmy, A. (2003) 'Mobility-Assisted Resolution of Queries in Large-Scale Mobile Sensor Networks,' *Computer Networks* special issue on Wireless Sensor Networks.

Herzberg, A., Jarecki, S., Krawczyk, H. and Yung, M., (1995) 'Proactive secret sharing or: How to cope with perpetual leakage,' proceedings of Crypto'95, pp. 339–352.

Hu, L. and Evans, D. (2003) 'Using Directional Antennae to Prevent Wormhole attacks,' 11th Network and Distributed System Security Symposium, 2003.

Hu, L., Perrig, A. and Johnson, D. B. (2003) 'Packet Leashes: A Defense Against Wormhole Attacks in Wireless Ad Hoc Networks,' INFOCOM 2003.

Hu, Y., Perrig, A. and Johnson, D. B. (2005) 'Ariadne: A Secure On-demand Routing Protocol for Ad Hoc Networks,' *Wireless Networks*, **11**, 21–38.

IEEE (2005a) Standard for Local and Metropolitan Area Networks – Specific requirements Part 15.1: Wireless Medium Access Control (MAC) and Physical Layer (PHY) Specifications for Wireless Personal Area Networks (WPANs).

IEEE (2005b) Standard for Local and Metropolitan Area Networks Part 16: Air Interface for Fixed Broadband Wireless Access Systems Amendment 2: Physical and Medium Access Control Layers for Combined Fixed and Mobile Operation in Licensed Bands.

IEEE-SA Standards Board (2003) 'IEEE Std. 802.15.4,' IEEE.

Ingemarsson, I., Tang, D. and Wong, C. (1982) 'A conference key distribution system,' *IEEE Transactions on Information Theory*, **28**(5), 714–720.

Intanagonwiwat, C., Govindan, R. and Estrin, D. (2000) 'Directed Diffusion: A Scalable and Robust Communication Paradigm for Sensor Networks,' ACM MobiCom '2000.

ITU-T (1991) Recommendation X.800, 'Security Architecture for Open System Interconnection for CCITT Applications'.

Joint Chiefs of Staff (2007) '*Electronic Warfare*,' Joint Publication 3-13.1, 'http:// www.fas.org/irp/doddir/dod/jp3-13-1.pdf,' January.

Joshi, D., Namuduri, K. and Pendse, R. (2005) 'Secure, Redundant, and Fully Distributed Key Management Scheme for Mobile Ad Hoc Networks: An Analysis,' *EURASIP Journal on Wireless Communication and Networking*, **5**(4), 579–589.

Jung, E. S. and Vaidya, N. H. (2002) 'A Power Control MAC Protocol for Ad Hoc Networks,' ACM MobiCom, September 2002.

Karl, H. and Willig, A. (2005) *Protocols and Architectures for Wireless Sensor Networks*, John Wiley & Sons, Ltd,.

Karlidere, T. and Cayirci, E. (2006) 'A MAC Protocol for Tactical Underwater Surveillance Networks,' MILCOM'06, Washington.

Karlof, C. and Wagner, D. (2003) 'Secure Routing in Wireless Sensor Networks: Attacks and Countermeasures,' *Ad Hoc and Sensor Networks*, **1**, 293–315.

Karn, P. (1990) 'MACA – a new channel access method for packet radio,' ninth ARRL Computing Networking Conference, September 1990, pp. 134–140.

Khalili, A., Katz, J. and Arbaugh, W. A. (2003) 'Towards secure key distribution in truly ad-hoc networks,' proceedings of the IEEE Workshop on Security and Assurance in Ad hoc Networks.

Klima, V. (2006) 'Tunnels in Hash Functions: MD5 Collisions Within a Minute,' *Cryptology* ePrint Archive, Report 2006/105.

Komerling, O. and Kuhn, M. G. (1999) 'Design principles for tamper resistant smartcard processors,' USENIX Workshop on Smartcard Technology, May.

Kong, J., Zerfos, P., Luo, H., Lu, S. and Zhang, L. (2001) 'Providing robust and ubiquitous security support for mobile ad-hoc networks,' proceedings of the ninth International Conference on Network Protocols (ICNP'01), pp. 251–260.

Kumar, S. Raghavan, V. S. and Deng, J. (2006) 'Medium access control protocols for ad hoc wireless networks: a survey,' *Ad Hoc and Sensor Networks*, **4**, 326–358.

Law, Y. W. (2005) 'Key Management and Link-Layer Security of Wireless Sensor Networks, Energy-efficient Attack and Defence,' PhD Thesis, CTIT PhD-thesis Series, Series number: 1381–3617, CTIT Number: 05-75.

Levine, J. (1999) 'Time synchronization over the Internet using an adaptive frequency locked loop,' *IEEE Transactions on Ultrasonics, Ferroelectronics, and Frequency Control*, **46**(4), 888–896.

Li, Z., Trappe, W., Zhang, Y. and Nath, B. (2005) 'Robust Statistical Methods for Securing Wireless Localization in Sensor Networks,' International Conference on Information Processing in Sensor Networks (IPSN'05).

Lin, X., Kwok, Y. and Lau, V.K.N. (2003) 'Power Control for IEEE 802.11 Ad Hoc Networks: Issues and a New Algorithm', International Conference on Parallel Processing (ICPP).

Liu, D. and Ning, P. (2003a) 'Establishing Pairwise Keys in Distributed Sensor Networks,' proceedings of CCS'03.

Liu, D. and Ning, P. (2003b) 'Efficient distribution of key chain commitments for broadcast authentication in distributed sensor networks,' proceedings of the 10th Annual Network and Distributed System Security Symposium, February 2003, pp. 263–276.

Liu, D., Ning, P. and Du, W. (2005a) 'Detecting Malicious Beacon Nodes for Securing Location Discovery in Wireless Sensor Networks,' 25th IEEE International Conference on Distributed Computing Systems (ICDCS'05), pp 609–619.

Liu, D., Ning, P. and Du, W. (2005b) 'Attack Resistant Location Estimation in Sensor Networks,' 4th International Conference on Information Processing in Sensor Networks (IPSN'05), pp 99–106.

Lynn, B. (2002) 'Authenticated identity-based encryption,' *Cryptology* ePrint Archive, iacr (International Association for Cryptological Research).

Marti, S., Giuli, T. J., Lai, K. and Baker, M. (2000) 'Mitigating routing misbehavior in mobile ad hoc networks,' MobiCom 2000.

McGrew, D. A. and Sherman, A. T. (2003) 'Key Establishment in Large Dynamic Groups Using One-Way Function Trees,' *IEEE Transactions on Software Engineering*, **29**(5), 444–458.

Meguerdichian, S. *et al.* (2001) 'Localized Algorithms in Wireless Ad hoc Networks: Location Discovery and Sensor Exposure,' *Proceedings of the IEEE/ACM Workshop on Mobile Ad Hoc Networking and Computing (MobiHOC2001)*.

Merwe, J. V. D., Dawoud, D. and McDonald, S. (2005) 'A Survey on Peer-to-Peer Key Management for Military Type Mobile Ad Hoc Networks,' Military Information and Communications Symposium of South Africa – MICSSA.

Mills, D. L. (1994) 'Internet time synchronization: the network time protocol,' in *Global States and Time in Distributed Systems*, IEEE Computer Society Press.

Mills, D. L. (1998) 'Adaptive hybrid clock discipline algorithm for the network time protocol,' *IEEE Transactions on Networking*, **6**(5), 505–514.

Mishra, A., Nadkarni, K. and Patcha, A. (2004) 'Intrusion Detection in Wireless Ad Hoc Networks', *IEEE Wireless Communications*, **11**(1), 48–60.

Newsome, J., Shi, E., Song, D. and Perrig, A. (2004) 'The sybil attack in sensor networks: analysis & defenses,' third International Symposium on Information Processing in Sensor Networks.

Niculescu, D. and Nath, B. (2003) 'Localized Positioning in Ad Hoc Networks,' IEEE SNPA 2003, pp. 42–50.

NIST SP-800-97 – Guide to 802.11i – Establishing Robust Security Networks.

Obraczka, K. (1998) 'Multicast transport protocols: a survey and taxonomy,' *IEEE Communications Magazine*, January, pp 94–102.

Onel, T., Ersoy, C. and Cayirci, E. (2004) 'Handoff Techniques for the VCL Based Mobile Subsystem of the Next Generation Tactical Communication Systems', *Computer Networks*, **46**(5), 695–708.

Ozturk, C., Zhang, Y. and Trappe, W. (2004) 'Source location privacy in energy constraint sensor network routing,' second ACM Workshop on Security of Ad Hoc and Sensor Networks.

Pan, J., Cai, L., Shen, X. and Mark, J. W. (2007) 'Identity based secure collaboration in wireless ad hoc networks,' *Computer Networks*, **51**, 853–865.

Papadimitratos, P. and Haas, Z. J. (2003) 'Secure link state routing for mobile ad hoc networks,' International Symposium on Applications and the Internet.

Parno, B., Gaustad, E., Luk, M. and Perrig, A. (2006) 'Secure Sensor Network Routing: A Clean State Approach,' CoNEXT 2006.

Patwari, N., Hero, A. O., Perkins, M., Correal, N. S. and O'Dea, R. J. (2003) 'Relative location estimation in wireless sensor networks,' *IEEE Transactions on Signal Processing.*

Pedersen, T. (1991) 'Non-interactive and information-theoretic secure verifiable secret sharing,' proceedings of CRYPTO'91, pp. 129–140.

Perrig, A., Canetti, R., Tygar, J. D. and Song, D. (2000a) 'Efficient Authentication and Signaling of Multicast Streams over Lossy Channels,' proceedings of IEEE Symposium on Research in Security and Privacy, pp. 56–73.

Perrig, A., Canetti, R. and Whillock, B. (2000b) 'TESLA: Multicast Source Authentication Transform Specification,' in draft-ietf-msec-tesla-spec-00.

Perrig, A., Canetti, R., Song, D., Tygar, J.D. and Briscoe, B. (2003) 'TESLA: Multicast Source Authentication Transform Introduction,' in draft-ietf-msec-tesla-intro-03.

Perrig, A., Song, D. and Tygar, J. D. (2001a) 'ELK, a new Protocol for Efficient Large-Group Key Distribution,' proceedings of IEEE Symposium on Security and Privacy.

Perrig, A., Szewczyk, R., Wen, V., Culler, D. and Tygar, J. D. (2001b) 'SPINS: Security Protocols for Sensor Networks,' MobiCom 2001.

Perrig, A., Szewczyk, R., Tygar, J. D., Wen, V. and Culler, D. (2002) 'SPINS: Security Protocols for Sensor Networks,' *Wireless Networks*, **8**(5), 521–534.

Perrig, A. and Tygar, J. D. (2003) *Secure Broadcast Communication in Wired and Wireless Networks*, Kluwer Academic Publishers.

Pietro, R. D., Mancini, L. V. and Jajodia, S. (2002) 'Efficient and Secure Key Management for Wireless Mobile Communications,' proceedings of POMC'02.

Poovendran, R. and Baras, J. S. (1999) 'An Information Theoretic Analysis of Root-Tree Based Secure Multicast Key Distribution Schemes,' *LNCS*, **1666**, 624–638.

Pottie, G. J. and Kaiser, W. J. (2000) 'Wireless Integrated Network Sensors,' *Communications of the ACM*, **43**(5), 551–558.

Priyantha, N.B., Chakraborty, A. and Balakrishnan, H. (2000) The cricket location support system, proceedings of the Sixth Annual ACM International Conference on Mobile Computing and Networking (MobiCom), August 2000.

Puzar, M., Andersson, J., Plagemann, T. and Roudier, Y. (2005) 'SKiMPy: A Simple Key Management Protocol for MANETs in Emergency and Rescue Operations,' proceedings of ESAS'05.

Rafaeli, S. and Hutchison, D. (2003) 'A Survey of Key Management for Secure Group Communication,' *ACM Computing Surveys*, **35**(3), 309–329.

Rafaeli, S., Mathy, L. and Hutchison, D. (2001) 'EHBT: An efficient protocol for group key management,' *LNCS*, **2233**,159–171.

RFC 2828 Internet Security Glossary May 2000.

Rhee, K. H., Park, Y. H. and Tsudik, G. (2004) 'An Architecture for Key Management in Hierarchical Mobile Ad-hoc Networks,' *Journal of Communications and Networks*, **6**(2), 156–162.

Rhee, K. H., Park, Y. H. and Tsudik, G. (2005) 'A Group Key Management Architecture for Mobile Ad-hoc Wireless Networks,' *Journal of Information Science and Engineering*, **21**(2), 415–428.

Rivest, R., Shamir, A. and Adleman, L. (1978) 'A Method for Obtaining Digital Signatures and Public Key Cryptosystems,' *Communications of the ACM*, February 1978.

Roman, R., Zhou, J. and Lopez, J. (2005) 'On the Security of Wireless Sensor Networks', proceedings of the 2005 ICCSA Workshop on Internet Communications Security, Singapore, *LNCS*, **3482**, 681–690.

Roman, R., Zhou, J. and Lopez, J. (2006) 'Applying Intrusion Detection Systems to Wireless Sensor Networks', IEEE Consumer Communications & Networking Conference (CCNC 2006), Las Vegas (EEUU), January 2006.

Royer, E. M. and Toh, C. (1999) 'A Review of Current Routing Protocols for Ad Hoc Mobile Wireless Networks,' *IEEE Personal Communications Magazine*, **6**(2), pp. 46–55.

Sadagopan, N., Krishnamachari, B. and Helmy, A. (2003) 'The Acquire Mechanism for Efficient Querying in Sensor Networks,' *Ad Hoc Networks.*

Salem, N. B., Buttyan, N., Hubaux, J. and Jakobsson, M. (2003) 'A Charging and Rewarding Scheme for Packet Forwarding in Multi-hop Cellular Networks,' MobiHoc 2003.

Sankarasubramaniam, Y., Akan, O. B. and Akyildiz, I. F. (2003) 'ESRT: Event-to-Sink Reliable Transport in Wireless Sensor Networks,' ACM Mobihoc'03.

Sanzgiri, K., Levine, B. N., Shields, C., Dahill, B. and Belding-Royer, E. M. (2002) 'A Secure Routing Protocol for Ad Hoc Networks,' International Conference on Network Protocols (ICNP).

Sasse, A., Brostoff, S. and Weirich, D. (2001) 'Transforming the "weakest link" – a human / computer interaction approach to usable and effective security', *BT Technology Journal*, **19**(3), 122–131.

Savvides, A., Han, C. and Srivastava, M. (2001) 'Dynamic fine grained localization in ad-hoc networks of sensors,' proceedings of MobiCom'01.

Schurgers, C., Kulkarni, G. and Srivastava, M. B. (2002) 'Distributed On-Demand Address Assignment in Wireless Sensor Networks,' *IEEE Transactions on Parallel and Distributed Systems*, **13**(10).

Selcuk, A., McCubbin, C. and Sidhu, D. (2000) 'Probabilistic Optimization of LKH-based Multicast Key Distribution Schemes,' IETF Internet-Draft, January, 2000, draft-selcuk-probabilistic-lkh-00.txt.

Shamir, A. (1979) 'How to share a secret,' *Communications of the ACM*, **22**, 612–613.

Shamir, A. (1984) 'Identity-based cryptosystems and signature schemes,' proceedings of CRYPTO '84, pp. 47–53.

Shen, C., Srisathapornphat, C. and Jaikaeo, C. (2001) 'Sensor Information Networking Architecture and Applications,' *IEEE Personal Communications*, **8**(4), 52–59.

Singh, S. and Raghavendra, C. S. (1998) 'PAMAS: Power Aware Multi-Access protocol with Signaling for Ad Hoc Networks,' *ACM Computer Communications Review*, July.

Smailagic, A., Siewiorek, D.P., Anhalt, J., Kogan, D. and Wang, Y. (2001) 'Location sensing and privacy in a context aware computing environment', proceedings of the International Conference on Pervasive Computing, pp. 15–23.

Sohrabi, K., Gao, J., Ailawadhi, V. and Pottie, G. (1999) 'A self-organizing sensor network,' proceedings of the 37th Allerton Conference on Communication, Control, and Computing, Monticello, Illinois, September.

Staddon, J., Miner, S., Franklin, M., Balfanz, D., Malkin, M. and Deam, D. (2002) 'Self-Healing Key Distribution with Revocation,' proceedings of the IEEE Symposium on Security and Privacy.

Stallings, W. (2000) *Data and Computer Communications*, sixth edition, Prentice Hall, Englewood Cliffs, New Jersey, USA.

Stallings, W. (2003) *Network Security Essentials*, Prentice Hall, Englewood Cliffs, New Jersey.

Stann, F. and Wagner, J. (2003) 'RMST: Reliable Data Transport in Sensor Networks,' IEEE SNPA 2003, pp. 102–112.

Steiner, M., Tsudik, G. and Waidner, M. (1998) 'CLIQUES: A new approach to Group Key Agreement,' proceedings of ICDCS'98.

Steiner, M., Tsudik, G. and Waidner, M. (2000) 'Key agreement in dynamic peer groups,' *IEEE Transactions on Parallel and Distributed Systems*, **11**(8), 769–780.

Subramanian, L. and Katz, R. (2000) 'An Architecture for Building Self-Configurable Systems,' *proceedings of theIEEE/ACM Workshop on Mobile Ad Hoc Networking and Computing (MobiHOC 2000)*, Boston, August 2000.

Tannenbaum, A. S. (2003) *Computer Networks*, Prentice Hall, Englewood Cliffs, New Jersey.

Tezcan, N., Cayirci, E. and Caglayan, U. (2004) 'End-to-end Reliable Event Transfer in Wireless Sensor Networks', PIMRC'2004.

Wallner, D., Harder, E. and Agee, R. (1999) 'Key Management for Multicast: Issues and Architectures,' IETF RFC 2627.

Walters, J. P., Liang, Z., Shi, W. and Chaudhary, V. (2006) 'Wireless sensor network security: a survey,' in *Security in Distributed, Grid and Pervasive Computing*, Y. Xiao (Ed.), Auerbach Publications/CRC Press..

Wan, C.-Y., Campbell, A. T. and Krishnamurty, L. (2003) 'PSFQ: A Reliable Transport Protocol for Wireless Sensor Networks,' ACM WSNA'02, Atlanta.

Wang, Y. (2005) 'Efficient Identity-Based and Authenticated Key Agreement Protocol,' *Cryptology* eprint Archive, Report 2005/108.

Waters, B. (2005) 'Efficient Identity-Based Encryption Without Random Oracles,' proceedings of Eurocrypt 2005.

Wi-Fi Alliance (2003) '*Wi-Fi Protected Access: Strong, standard-based interoperability security for today's Wi-Fi networks*'.

Winjum, E., Hegland, A. M., Spilling, P. and Kure, O. (2005) 'A Performance Evaluation of Security Schemes proposed for the OLSR Protocol,' proceedings of MILCOM'05.

Wong, C. K., Gouda, M. and Lam, S. S. (1998) 'Secure Group Communications Using Key Graphs,' proceedings of the SIGCOMM '98.

Wong, C. K. and Lam, S. S. (2000) 'Keystone: A Group Key Management Service,' proceedings of ICT 2000.

Wood, A. and Stankovic, J.A. (2005) 'A Taxonomy for Denial-of-Service Attacks in Wireless Sensor Networks', in *Handbook of Sensor Networks: Compact Wireless and Wired Sensing Systems*, CRC Press, pp. 32:1–20.

Wood, A. D., Fang, L., Stankovic, J. A. and He, T. (2006) 'SIGF: A family of configurable, secure routing protocols for wireless sensor networks,' ACM SASN.

Wu, B., Wu, J., Fernandez, E. B. and Magliveras, S. (2005) 'Secure and Efficient Key Management in Mobile Ad Hoc Networks,' proceedings of IPDPS'05.

Xi, Y., Schwiebert, L. and Shi, W. (2006) 'Preserving privacy in monitoring based wireless sensor networks,' second International Workshop on Security in Systems and Networks (SSN'06).

Ye, W., Heidemann, J. and Estrin, D. (2004) 'Medium Access Control with Coordinated, Adaptive Sleeping for Wireless Sensor Networks,' *IEEE/ACM Transactions on Networking*, **12**(3), 493–506.

Yi, S. and Kravets, R. (2002a) '*Key Management for Heterogeneous Ad Hoc Wireless Networks*,' University of Illinois at Urbana-Champaign.

Yi, S. and Kravets, R. (2002b) '*MOCA: MObile Certificate Authority for Wireless Ad Hoc Networks*,' Report No. UIUCDCS-R-2004-2502,UILU-ENG-2004-1805, University of Illinois at Urbana-Champaign.

Yi, S. and Kravets, R. (2004) 'Composite Key Management for Ad Hoc Networks,' proceedings of Mobiquitous'04.

Yu, B. and Xiao, B. (2006) '*Detecting selective forwarding attacks in wireless sensor networks*,' IEEE.

Zapata, M. G. and Asokan, N. (2002) 'Securing Ad Hoc Routing Protocols,' WiSe.

Zhang, Y., Lee, W. and Huang, Y. (2003) 'Intrusion Detection Techniques for Mobile Wireless Networks', *Wireless Networks Journal* (ACM WINET), **9**(5), September.

Zhou, L. and Haas, Z. J. (1999) 'Securing ad hoc networks,' *IEEE Network Magazine*, **13**(6), 24–30.

Zhu, S., Xu, S., Setia, S. and Jajodia, S. (2003a) 'Establishing pair-wise keys for secure communication in ad hoc networks: A probabilistic approach,' proceedings of the 11th IEEE International Conference on Network Protocols (ICNP'03), pp. 326–335.

Zhu, S., Setia, S. and Jajodia, S. (2003b) 'LEAP: Efficient Security Mechanisms for Large-Scale Distributed Sensor Networks,' proceedings of CSS'03.

Zhu, S., Setia, S., Xu, S. and Jajodia, S. (2004) 'GKMPAN: An Efficient Group Rekeying Scheme for Secure Multicast in Ad-Hoc Networks,' proceedings of Mobiquitous'04.

Zhu, B., Bao, F., Deng, R. H., Kankanhalli, M. S. and Wang, G. (2005) 'Efficient and robust key management for large mobile ad hoc networks,' *Computer Networks*, **48**(4), 657–682.

ZigBee Alliance (2004) 'ZigBee Standard, version 1.'

Zimmermann, P. (1994) *PGP User's Guide*, MIT Press.

Index